Werner Kunz

Do Species Exist?

Related Titles

Wiley, E. O., Lieberman, B. S.

Phylogenetics

Theory and Practice of Phylogenetic Systematics

2011

ISBN: 978-0-470-90596-8

Ladle, R. J., Whittaker, R. J. (eds.)

Conservation Biogeography

2011

ISBN: 978-1-4443-3504-0

Sarkar, S., Plutynski, A. (eds.)

A Companion to the Philosophy of Biology

2010

ISBN: 978-1-4443-3785-3

Knowles, L. L., Kubatko, L. S. (eds.)

Estimating Species Trees

Practical and Theoretical Aspects

2010

ISBN: 978-0-470-52685-9

Ayala, F. J., Arp, R. (eds.)

Contemporary Debates in Philosophy of Biology

2009

ISBN: 978-1-4051-5999-9

Rosenberg, A., Arp, R. (eds.)

Philosophy of Biology

An Anthology

2009

ISBN: 978-1-4051-8316-1

Ruse, M.

Charles Darwin

2008

ISBN: 978-1-4051-4913-6

Werner Kunz

Do Species Exist?

Principles of Taxonomic Classification

The Author

Prof. Dr. Werner Kunz
Heinrich-Heine-University
Institute for Genetics
Universitätsstraße 1
40225 Düsseldorf
Germany

Cover

Diversity of Species in Central American Rain Forest (photomontage). In the center Resplendent Quetzal (*Pharomachrus mocinno*), on its left on the top Chestnut-mandibled Toucan (*Ramphastos swainsonii*) and the smaller green Emerald Toucanet below (*Aulacorhynchus prasinus*). In the bottom left corner a flying Swallowtail (*Parides lycimenes*). On the right side of the figure from top to bottom Red-lored Parrot (*Amazona autumnalis*), Squirrel Monkey (*Saimiri oerstedii*), Fiery-billed Aracari (*Pteroglossus frantzii*), a flying Heliconiid butterfly (*Dryas julia*), Roadside Hawk (*Buteo magnirostris*), White-headed Capuchin (*Cebus capucinus*), Ferruginous Pygmy Owl (*Glaucidium brasilianum*) and at the bottom in the sky three Blues of genus *Strymon*.

Library of Congress Card No.: applied for

British Library Cataloguing-in-Publication Data
A catalogue record for this book is available from the British Library.

Bibliographic information published by the Deutsche Nationalbibliothek
The Deutsche Nationalbibliothek lists this publication in the Deutsche Nationalbibliografie; detailed bibliographic data are available on the Internet at http://dnb.d-nb.de.

Wiley-Blackwell is an imprint of John Wiley & Sons, formed by the merger of Wiley's global Scientific, Technical, and Medical business with Blackwell Publishing.

Cover: Formgeber, Eppelheim
Typesetting: Thomson Digital, Noida, India
Printing & Binding: Markono Print Media Pte Ltd, Singapore

Print ISBN: 978-3-527-33207-6
ePDF ISBN: 978-3-527-66425-2
ePub ISBN: 978-3-527-66426-9
mobi ISBN: 978-3-527-66427-6
oBook ISBN: 978-3-527-66428-3

Contents

Foreword

Do species exist?

For Ernst Haeckel, in the midst of nineteenth century, the point was quite easy: In the beginning, there was a plant and an animal. Evolution started extending and varying their principal organizations. Accordingly, species were regarded to be just notes given by man to the continuous flow of evolving life. Classifying man looked for these entities to allow him a qualification of the ongoing diversification of life. Taxonomy, thus, supporting such an evolutionary biology should form out something like a phylogenetic systematics that may outline the quality of that continuing process of evolution in its details. Haeckel did not succeed to from out such a new taxonomy even though he was one of the great taxonomists of the late nineteenth century. In an evolutionary perspective, addressing that continuous flow of evolution, species, thus, could be described only as categories formed by the human mind. With the introduction of population genetics into Darwinism, the situation changed. Genetics allowed to address a certain base of heredity. Accordingly, the continuous flow of evolving forms could be outlined in a much more substantiated way, allowing to address how far genes were established, varied, or deleted in the course of generations. In fact, more and more extending insights into the functional organization of life forms established new ideas about life organization, in general. Thus, today, the situation of taxonomy is far less clear than it was at Haeckel's time. Not only that the last unified common ancestor had to be a prokaryote or we had to address mushrooms and eventually such prokaryotes and some others of such organizations as principally differing basic types of life. We have to integrate morphology, population biology, and molecular genetics and their different accesses toward a species concept. Thus, we have to address the question of what a species really is, again. And we have to clarify several problems: How we can evaluate biological diversity if species do not exist? How we can understand evolution if speciation is not really the motor of ongoing evolutionary development? And how we can compare the accesses of molecular biology, the analysis of palaeontology, cladistics, and morphological analyses directed toward the species concept in one run? In his book, Werner Kunz describes and analyzes all such various interpretations and concepts dealing with species. Thereby, he shows us that the different ideas of Neo-Darwinism, taxonomy, and genetics do not fit into each other. Even worse, he is aware that species have been formed out in various solutions within different evolutionary settings. A plant species is not directly comparable with an animal or mushroom species. In an evolutionary perspective, all these species are the outcome of differing evolutionary strategies. That is not true for their differentiation in taxonomic regard, but it is true in regard of what *species* mean in life. Thus far, a species not only might be characterized by a set of attributes allowing a reliable classification but may also offer principal differing materials for an ongoing evolution. There, species of plants, various animals, prokaryote species, and mushrooms may each react differently in evolution. Already, sexuality is organized in all these life forms in a different way. In any case, there is just something transferring a set of genes, certain morphological and functional specifications from one to the next generations. The modes by which this is being practised are different, however. What should be done in such a situation?

Werner Kunz did not offer a philosophical solution. He is presenting facts, and he is doing that in a comparative perspective. One point Kunz makes is that the taxonomists rely on Linnaeus who had a completely different view on nature from that we have today. For Linnaeus any species is part of creation. If nature is such a creation done by God, any entity in nature is reflecting an absolute ordering scheme. Systematic will outline this scheme. Thus, live forms a thought to be organized like the terms in a baroque encyclopedia. There, any term outlines a basic idea. Its true meaning is intelligible when the order, in which it is used, is made obvious. The underlying structure by which such ideas could be combined is the idea of a universal topic reflecting the concept God has had in mind in setting out his creation. To combine such a scheme with the Darwinian idea of a continuously varying world is not possible. Nevertheless, idealistic morphology in the start of twentieth century tried to do this. The result was a logical scheme adopted in principle by Willi Hennig. He formed out an abstract pattern that allowed a proper classification, but was not interested to integrate that view with a historical reconstruction of what actually happened in the course of evolution. Cladism adopted Hennig's idea, which is now forming the conceptual framing for evolutionary interpretation of DNA analyses. When such an idea of a logically consistent scheme is combined with evolutionary population biology, problems occur. Accordingly, in an attempt to combine such approaches, one has to address anew the question what a species really is. If species are individuals, evolutions will deal with them, resulting in new species. If species consist of populations, and if microevolution works on the level of such population, situations might become more difficult. What a species meant, cannot just be a taxonomic entity without any functional relevance in evolutionary biology. Would that be the case, we could not describe evolution as a process resulting in speciation. If species are actually something evolution worked on, then, however, species themselves (as structural units) might be entities that have been evolved as such ones within various evolutionary constraints. Accordingly, a species might address something different in plants, mushrooms, bacteria, viruses, and animals. On the other hand, what is meant by such different concepts regarding the functionality of the species? The resulting idea, describing evolutionary relevance of species within a certain evolutionary process, might differ from what a species meant for other life forms. A rodent will eat certain plants and will avoid poisonous ones. Thus, the species' concept somehow is actually valid for it irrespective of the different evolutionary dynamics of such a poisonous plant and its own species.

Everyone describing biodiversity has to encounter diversification on the species level. He has to think about what it means to be extinct. He may even understand that a species is formed by populations and he has to understand that a population is not a species. The problems that come out of all that are addressed by Werner Kunz. He is not offering a new philosophy. He is following the ideas in biology to their consequences. This allows an understanding of the various uses and the significance of species concepts. This enables to address a lot of relevant questions and to understand conceptual constraints in modern evolutionary biology. The result is a great book that should be read by anyone who wants to understand evolution, biodiversification, and the meaning of species in those.

Ernst-Haeckel-Haus, Jena, April 2012 *Olaf Breidbach*

Preface

What is water? You would not answer "water is wet" or "water is a liquid" because these are properties of the water, not definitions. The answer "water is wet" does not explain what water is. Water is a substance consisting of H_2O molecules, and the formula H_2O is exhaustively explained by physicists and chemists. It is well known what water is.

However, what is a tiger? If you would answer "a carnivorous animal which is big and has black stripes," you would describe only the properties of the tiger. You would not answer the question, what a tiger is. If you try to find the answer to the question what a tiger is, you will notice that the answer is very difficult, if not impossible to find.

The same becomes apparent if you ask the question "what is a species?" It is relatively easy to answer the question "what is a molecule?" A molecule is a group of atoms that are linked by chemical bonds, and atoms and chemical bonds are well defined by physicists and chemists.

Like a chemical molecule, a biological species is a group of organisms. But are the organisms of a species linked by bonds? If yes, the question arises what kind of bonds these are that hold the organisms of a species together. If there are no bonds, the question arises why the individuals of a species can form a group at all. If there are no groups, there are no species.

It is absolutely necessary to define the term group in taxonomy because there are several kinds of groups that organisms can form in nature. For example, migrating birds can form stable groups. You also can classify apes and monkeys including humans in groups according to their blood group alleles. But such groups of organisms are not species. What is it that makes a group of organisms to be a species? If you try to find the answer to the question what a species is, you will notice that the answer is very difficult, if not impossible to find. What is it that makes a tiger to be a member of the species tiger? And what is the species tiger?

If you consider all organisms that resemble each other in their traits to belong to a species, this view is immediately upset by the phenomenon of distinct trait differences between the two sexes of the same species that often exceed the species differences. If you consider all organisms to belong to a species that are able to crossbreed successfully, this view cannot be held out because several individuals of a species cannot crossbreed successfully for genetic or other reasons. If you consider all organisms to belong to a species whose genomes are similar in DNA sequence, this view will soon be corrected by the awareness that there exist phylogenetically young species that are genetically homogeneous as well as phylogenetically old species that are genetically heterogeneous. There may be larger genetic differences among the members of an evolutionary old species than among the members of different species if the species are evolutionary young. Evolutionary distance cannot be a reliable species criterion. At the end, all attempts to discriminate species from each other are blurred by the fact that all criteria for these discriminations also may apply to the individuals within a species. Intraspecific polymorphisms are a main obstacle for taxonomic classification.

It appears that, in contrast to chemical objects, taxonomic objects cannot be defined. What is it that makes this fundamental difference between chemical and biological objects? It is very remarkable that taxonomists in most cases abandon the species problem and are nevertheless quite able to work with species. How is it possible to do scientific work and to obtain reproducible results with objects without knowing what these objects are? Should it not be irritating or alarming to do research with objects that are not defined?

Despite several promises, the species problem is not solved, and it cannot be ignored. This book elucidates the inconsistencies and contradictions that stand behind the conventional species concepts. In this book, it is emphasized repeatedly that it is a doubtful, if not an unscientific, way to practice taxonomic classifications while ignoring the foundations of the species problem. Furthermore, beyond these theoretic considerations, uncertainties and ambiguities of the species problem have considerable impact on the strategies of legislation in biodiversity conservation politics.

Düsseldorf, April 2012 *Werner Kunz*

Color Plates

Do Species Exist? Principles of Taxonomic Classification, First Edition. Werner Kunz.

Color Plate 1. Multivoltine Butterflies (photomontage)
Several butterfly species can be univoltine or bivoltine or even trivoltine. These are different morphs of the same species that differ genetically. Univoltine are individuals which produce only one imaginal generation per year; they live in the north. Bi- or trivoltine individuals generate two or three imaginal generations per year; in spring, early and late summer. They live in more southern regions. In Europe, well-known examples are the Scarce Swallowtail *Iphiclides podalirius* (top), the Common Swallowtail *Papilio machaon* (left), the Common Blue *Polyommatus icarus* (bottom left) and the Peacock butterfly *Inachis io* (bottom right).

Color Plate 2. Batesian Mimicry (photomontage)
Similarity is not kinship. Heliconiid and Ithomiid butterflies are bad-tasting and therefore avoided as a prey by birds and other predators. Representatives of completely different families of butterflies imitate the shapes and color patterns of the unpalatable species to be protected, although they are not unpalatable. Each of the four groups presents a Central American Heliconiid or Ithomiid together with their imitators. *1 Heliconius ismenius* together with *2 Eresia eutropia (Nymphalidae)*; *3* the Heliconiid *Eueides isabella* together with *4 Dismorphia orise* (*Pieridae*); *5* the Ithomiid *Ithomia heraldica* together with *6 Actinote anteas (Acraeidae)*; *7 Eueides isabella* together with *8 Dismorphia amphiona* (*Pieridae*).

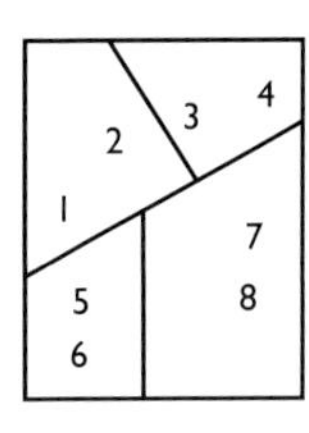

Scheme for Color Plate 2

Color Plate 3. Müllerian Mimicry (photomontage)
Similarity is not kinship. Different Heliconiid and Ithomiid species in Central America resemble each other in shape and color pattern, although they belong to two different butterfly families. Each of the four groups presents a Heliconiid together with Ithomiids. *1 Heliconius hecale* together with *2 Mechanitis lysimnia*; *3 Eueides isabella* together with *4 Hypothyris lycaste*; *5 Heliconius ismenius* together with *6 Napeogenes tolosa* and *7 Godyris zavaleta*; *8 Heliconius hecale* together with *9* and *10 Melinaea scylax* and *11 Mechanitis polymnia*.

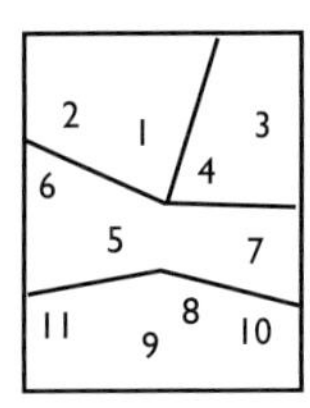

Scheme for Color Plate 3

Color Plate 4. Cryptic Species and Mimicry Morphs (photomontage)
Six different species of European Burnet moths (family *Zygaenidae*) are very similar in phenotype. They all are black with brilliant red spots. From left to right: *Zygaena filipendulae*, *Z. ephialtes* (photo: Jochen Rodenkirchen), *Z. transalpina*, *Z. viciae* (on top of blade of grass), *Z. lonicerae* (on yellow Birdsfoot Trefoil flower *Lotus corniculatus*) and *Z. angelicae* (on top of purple Oregano flower *Origanum vulgare*). *Z. ephialtes*, however, is also found in a completely different morph that has white and yellow spots (two examples bottom center and bottom right). This dark morph resembles the nine-spotted moth (*Amata phegea*), a moth belonging to the family *Arctiidae* (bottom left).

Color Plate 5. Sexual Dimorphism (photomontage)
Similarity is not kinship. Differences in traits between the morphs within a species can be much greater than species differences. Here sexual differences between males and females of Central American Tanagers are shown. *1* and *2* Cherrie's Tanager (*Ramphocelus costaricensis*) female and male; *3* and *4* Summer Tanager (*Piranga rubra*) male and female; *5* and *6* Passerini's Tanager (*Ramphocelus passerinii*) female and male.

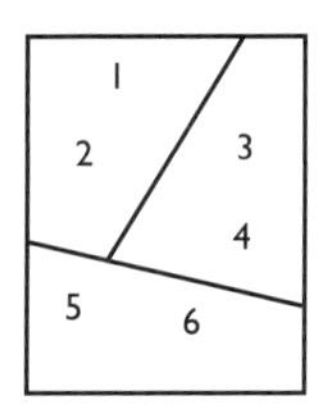

Scheme for Color Plate 5

Color Plate 6. Confusing Similarity between *Doxocopa* and *Adelpha* (photomontage) Similarity is not kinship. In Latin American rain forest, the females of *Doxocopa* species (*1* and *2*) resemble the butterflies of another Nymphalid genus: *Adelpha* (*3* – *5*). They don't resemble the males of their own species, which are shining brilliant blue (*6* – *8*). *1 Doxocopa excelsa* and *2 Doxocopa laurentia*; *3 Adelpha zea*, *4 Adelpha basiloides* and *5 Adelpha cytherea*; *6* and *8 Doxocopa laurentia*, together with *7 Doxocopa clothilda*.

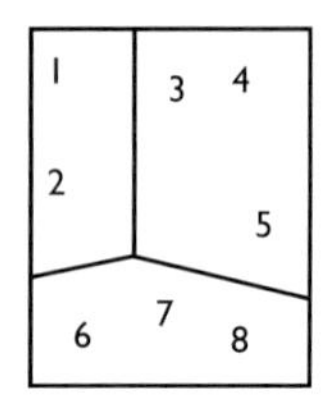

Scheme for Color Plate 6

Color Plate 7. Cryptic Species and an abnormal Mutant (photomontage)
Several European Fritillaries resemble each other largely, and it is very hard to determine their species membership. However, a rare mutant of *Melitaea athalia* is aberrant in its color pattern and deviates from the other members of its own species much more than the members of different species differ from each other. Above center: two Nickerl's Fritillaries *Melitaea aurelia*; above right: Assmann's Fritillary *M. britomartis*; center below the Nickerl's Fritillaries: two normal Heath Fritillaries *M. athalia*, and bottom left a mutant of *M. athalia*.

Color Plate 8. Water Frogs are "Kleptospecies" (photomontage)
The Water Frog *Pelophylax esculenta* (top and center) is a hybrid species between the Marsh Frog *P. ridibunda* (bottom right) and the Pool Frog *P. lessonae* (bottom left; photo: Benny Trapp). Whereas the somatic cells of the Water Frog contain the genomes of both parental species, one of the genomes is completely eliminated in the germinal cell line in the Water Frog prior to meiosis. As a consequence, the genes of one of the two parental species cannot be passed on to further generations. Therefore, the Water Frog "steals" one of its genomes from a foreign species and is termed a "kleptospecies" for this reason.

Introduction

In one of his most famous works, "Die Harzreise" ("The Harz Journey") (Heine, 1826), the German poet Heinrich Heine found a beautiful flower on the Brocken, the highest peak in the German Harz mountains. Tourists were standing nearby in considerable numbers, and they all wanted to know the name of the flower. Heine expressed particular aversion to this demand and wrote:

"It always annoys me to see that God's dear flowers have been divided into castes, just like ourselves, and according to similar external features like differing stamens. If there has to be classification, people should follow the suggestion made by Theophrastus, who wanted to classify flowers in a more spiritual manner, that is, by scent. As for me, I have my own system of natural history, according to which I classify everything as eatable or uneatable" (The Harz Journey; English translation Heine) (Heine, 2006).

This almost two-century-old assessment of taxonomy is not as absurd as it might at first appear. Even today, taxonomic classifications are based on several subjective standards of valuation that do not withstand a thorough theoretical test. There are several reasons for this problem (Mallet, 1995). One reason is that, even today, no theory in taxonomy can determine and prescribe which traits may be used in taxonomic classification and which may not.

No binding rules exist for the classification of individuals into species, nor is it clear whether species exist at all. Notwithstanding the title of his famous work "On the Origin of Species," Charles Darwin did not believe in the existence of species (Chapter 2). He took species for constructed units, defined to achieve a comfortable ordering of living beings that would not even exist in the absence of human principles of classification. Many recent authors treat the biological species in precisely the same way.

Several taxonomists agree that a definition of the term "species" will never be possible. Indeed, they state that this issue is merely an "academic" question and that it is not meaningful for a scientist to devote time to such a problem. To these taxonomists, the expectation that the nature of species can be understood represents an illusory goal that cannot be achieved.

As an alternative, they restrict taxonomy to the simple goal of identification. To work with species and to understand the function of a species and its ecological role, it

Do Species Exist? Principles of Taxonomic Classification, First Edition. Werner Kunz.

would be sufficient to identify the units of biodiversity without knowing what these units are. Questions such as "what is a tiger?" or "what makes a tiger a tiger?" are considered to be senseless questions that hamper science.

This type of reductionism, "taxonomy is diagnosis or identification," generated the recent technology of barcoding, a method that is evaluated as "the future of taxonomy" for several reasons. Supported by substantial funding, assisted by very successful public relations and preceded by the belief that this approach represents genuine "high-tech taxonomy," the barcoding-technology approach to taxonomy has initiated a triumphal procession (Chapter 4).

However, it seems to be forgotten that identification cannot be the ultimate goal of taxonomy. What result is ultimately obtained if groups of individuals are identified? The use of diagnostic tools can allow the identification of a number of cat-like animals as tigers or as lions. However, a number of individuals can also be identified as males or as females. Furthermore, a number of individuals can be identified and distinguished by differences in their blood groups. The results obtained by these three approaches to identification are certainly very different, and the identification procedures per se do not distinguish between intraspecific polymorphisms and species differences.

The attempts to identify animal or plant groups do not achieve the final goal of demonstrating that these groups are species. It does not suffice that the groups can merely be distinguished from each other. Why are the different sexes not different species? Neither the simplest conservative identification techniques nor the most modern molecular techniques can determine whether two clearly distinguishable groups are species. This consideration shows that the idea of species has additional significance. It is the goal of this book to elucidate the true nature of species. Species are not simply groups of individuals that can be distinguished. Species are something else entirely.

Educated by a number of identification guides or field guides that are available for most groups of animals or plants, we are misled to believe that individuals that clearly differ in traits must belong to different species. If a goose in Eurasia has a uniformly pink-to-orange beak and pink feet, it must be a Greylag Goose (*Anser anser*). However, if it has an orange beak with black margins together with orange feet, then it must be a Bean Goose (*Anser fabalis*). Nevertheless, these differences are not exhibited by each individual Greylag Goose or Bean Goose because mutants occur. Hence, why are those mutants still members of the species? Why are they not different species? Indeed, of what help are identifying traits for the understanding of the species?

Linnaeus stated that certain traits are essential to the species. A particular member of the species must possess these traits, or else it would not belong to the species. However, Darwin stated that the particular traits found in the individuals of a species change over time. This principle means that no single trait can be the essence of a species. It is not possible that both authors can simultaneously be correct.

As a consequence of Darwin's theory of evolution, it was necessary to conduct a thorough revision of Linnaeus's view of the species. This revision was achieved by Poulton in 1903, and extended by Dobzhansky in 1937 and Mayr in 1942. These authors replaced the Linnaean typological view of a species by the concept of a reproductive community. This concept is based on mutual lateral gene exchange by sexual contact. Organism *A* belongs to the same species as organism *B* if any of its offspring does receive genes from organism *B*.

The concept of the species as a gene-flow community is a species concept based on mutual relational connections among the organisms and their offspring (Chapter 6). It is not a typological species concept based on trait similarities. The concept of the gene-flow community is not easy to understand. It contradicts a type of cognitive presetting in the human mind (Chapter 2), and most importantly, it is very difficult to use. Ultimately, it is inapplicable for use in an operational and pragmatic everyday taxonomy. However, the concept of the gene-flow community appears to be the only species concept that reflects an entity that exists as reality as a delimited group in nature. According to this species concept, the borders between the species (although penetrable) exist in nature; they do not result purely from human constructs used for the purpose of grouping individuals.

The fundamental disagreement between a species concept that is logically consistent and a species concept that is applicable in practice is the primary reason for the existence of a "species problem" that could not previously be resolved. This deep conflict has its roots in the incompatibility between the claim to classify biodiversity according to taxa, in the sense of Linnaeus, and the scientific fact, introduced to the world by Darwin, that the traits of organisms change over time (Chapter 2).

This book has a long history. The book's inception occurred almost twenty years ago, when I became aware that the biological phenomenon of multiple allelic polymorphism implies a serious problem for taxonomic classification. How can organisms be classified into different species if single organisms already differ in hundreds or even thousands of their traits? Doesn't this mean that there must exist two different types of traits? One type of trait serves to discriminate among individuals within the same species, whereas other types of traits must possess certain unique qualities to be suited for species discrimination. However, two such types of traits do not exist (Chapter 4). Accordingly, what difference separates individual differences and species differences?

This book addresses biologists and philosophers, although it is much more a biological than a philosophical book. During the long time of the progress of this book, I benefited greatly from Markus Werning (now University of Bochum), who taught me several basic elements of philosophy. I also thank Gerhard Schurz (Düsseldorf), who opened the door for me to enter the philosophic scientific community. A decisive role in the continuation of my efforts to bind taxonomy to philosophy was played by Hartmut Greven (Düsseldorf), who encouraged me not to give up. He eased the difficulty for me, as a geneticist and molecular biologist, to gain entry into the taxonomic scientific community by inviting me to give lectures and to publish preliminary papers on the species problem. I also thank Sebastian Löbner (Düsseldorf), who is a linguist, not a biologist, but his invitation to be a member of his research group on functional concepts had a great impact on the understanding of taxonomic class formation presented in this book. Finally, I thank Gregor Cicchetti and Andreas Sendtko for their decision to support the processing of this book by the Wiley-Blackwell publishing company.

A number of the color plates in this book would not have been created if I had not enjoyed several holidays on the Finca Hamadryas in Costa Rica, where Paul Gloor helped me to identify the butterfly species that I photographed there. The book was first completed in German language, and afterwards translated into English. The first versions of this translation were done by Christian Feige. The art work of the black and white graphs in the book was done by Karin Kiefer. The color plates are photo montages of my own outdoor photographs that are inserted into the correct habitat. The technical art work was done by Monika Dörkes. I also thank Albert Kaltenberg who was the "soul" of my computer whenever the machine did not follow my advices.

This book presents no novel scientific data, nor does it present new philosophical conclusions. The material on which the book is based has previously appeared in biological books and papers or in philosophical books and reviews. The novel feature of this book is that it combines the fields of biology and philosophy. In this book, philosophical reasoning is explained for biologists and applied to unsolved or controversially disputed taxonomic problems. This book will raise biologists' awareness of one of the most difficult problems of taxonomy, namely, how to arrange the existing diversity of living organisms into cohesive and delimited groups.

Remarkably, many taxonomists are not genuinely interested in the species problem, although they are affected. In contrast, philosophers are more engaged with the foundations of taxonomy, although they are not as directly affected in their daily research. It was Albert Einstein who once said: "Science without philosophy is blind, and philosophy without science is empty."

1
Are Species Constructs of the Human Mind?

In 1926, Reagan defined the species as a purely pragmatic principle of classification: "A species is what a good taxonomist says it is" (cited from Huxley, 1942). In 1996, Hawksworth did not see the biological species any differently: "Species are groups of individuals separated by heritable character discontinuities and which it is useful to give a name to" (cited from Heywood, 1998). Even today, more than twenty different species concepts are still practiced concurrently (Mayden, 1997). This observation shows either that *the* biological species does not exist or that the particular species concepts define something different from the one truly existing species.

Since Darwin and Wallace, it has not been possible to unite Linnaeus's taxonomic principle of classification into rigid classes with the theory of evolution. Simply consider the implications of the title of a famous publication by Alfred Russel Wallace "On the tendency of varieties to depart indefinitely from the original type" (Wallace, 1858). Does this title in itself not mean "There are no species?"

With these considerations in mind, it would now be consistent and simple to accept the reality that species are fictitious human constructs made to sort genuine biodiversity into manageable but artificial units. However, a large majority of field biologists, insect collectors and "tickers" and "twitchers" among the hundreds of thousands of bird watchers believe in the real existence of species. All of the modern field guides to the birds of Europe and the adjoining regions contain approximately 800 bird species. None of these books identify the species concept that was used to obtain this number. They do not explain whether the term "species" means morphotypes, ecotypes, reproductive communities or descent communities. Instead, the impression is conveyed that these 800 species exist in reality and that each species simultaneously satisfies the classification principles furnished by each species concept.

Of course, the field guides do contain disputed borderline cases, for example, the recently undertaken separation of the Balearic Shearwater (*Puffinus mauretanicus*) from the Yelkouan Shearwater (*P. yelkouan*) or the separation of the eastern Mediterranean Black-eared Wheatear (*Oenanthe melanoleuca*) from the western Mediterranean Black-eared Wheatear (*O. hispanica*) to give two distinct species. However, these are isolated incidents. In the main, the books convey the general

Do Species Exist? Principles of Taxonomic Classification, First Edition. Werner Kunz.

consensus that species exist without posing the question of the nature of species. Otherwise, no consistent field guide could appear on the market. Nevertheless, these apparently unambiguous species are not defined anywhere in the field guides. Except for observations that certain species diverge from each other genetically or that there are diagnostic-typological differences, the reader does not learn why particular varieties are delimited from each other as species.

Adherence to any species concept is never fully consistent. If the reproductive community, the classification according to apomorphies or the classification of equal-ranking kinship were actually taken seriously, then many animal and plant groups would be split much more deeply into separate units than current practice supports (Chapter 2). An unspoken agreement appears to sanction "generously" combining mosaic-like fragmented reproductive communities or nested cladistic bifurcations to construct inclusive species boundaries because this approach yields readily manageable units. In critical cases, pragmatism proves to be a highly dominant principle in taxonomy. Pragmatism determines taxonomy's direction, and consistent reasoning has only a marginal importance in taxonomy (Chapter 2).

The introduction to a remarkable review article by Martin L. Christoffersen titled "Cladistic taxonomy, phylogenetic systematics, and evolutionary ranking" in the journal *Systematic Biology* contains the following statement that could equally be an opening theme for the present book (Christoffersen, 1995):

> "The ancient discipline of biological taxonomy has been very slow to incorporate major shifts in world views... Impervious to the derision of scientists in the more glamorous fields of research, many taxonomists today simply take for granted secular traditions of describing and naming the diversity of nature. They may persist stoically for a lifetime in such a self-appointed descriptive role, avoiding theory, philosophy and explanation. Some of these taxonomists may venture intuitive classifications for their named groups but will often delegate to others the task of deriving evolutionary meanings from their proposals."

Of course, one can use the traits employed for identification to recognize particular species and to distinguish them from other species. However, this procedure already implies that these particular species do exist, and that one needs only to learn how to identify them. If there were no species, it would be meaningless to identify them. Moreover, if two groups of organisms were not different species, but instead were one and the same species, it would be meaningless to identify and distinguish them. This observation demonstrates that the process of defining a species must precede the process of identifying that species (Chapter 2). Taxonomy cannot defend its reputation as a serious science if it relies exclusively on species identification. More scientific than the diagnosis of a species is the "why" of a species (Mayr, 2000). It is not sufficient to identify two organisms belonging to two different species by their diagnostic traits. It is more scientific to be able to explain the reasons that the organisms belong to two different species.

There is an important difference between that which something is and that by which something can be identified. Two human beings are not brothers because they have similar traits, but because they have the same parents. Half a century ago, George Gaylord Simpson stated this difference as follows: "The well-known example of monozygotic twins is explanatory. . . Two individuals are not twins because they are similar but, quite the contrary, are similar because they are twins" (Simpson, 1961). Stated precisely, individual organisms do not belong to the same taxon because they are similar, but they are similar because they belong to the same taxon.

The anthropologist and psychologist Scott Atran stated resignedly: "Perhaps the species concept should be allowed to survive in science more as a regulative principle that enables the mind to establish a regular communication with the ambient environment than as an epistemic principle that guides the search for nomological truth" (Atran, 1999).

It appears that species are simply pragmatic principles of classification. Furthermore, the principles of classification are not the same in higher animals, for example, antelopes in Africa, and in more primitive animals, for example, rotifers. However, under these conditions, the species of different animal and plant taxa are not mutually comparable. It would be meaningless to contrast the species richness of certain beetle families (*Coleoptera*) with the species poverty of certain families of frogs (*Anura*). Nevertheless, such comparisons are made.

Taxonomy pursues the intention of classifying organisms according to personal standards. In contrast, scientific correlations, as they nomologically exist in nature, are a different matter. To research such correlations serves a different objective and disagrees with taxonomy's goal of forming a stable classification (see Section "The constant change in evolution and the quest of taxonomy for fixed classes: can these be compatible?" in Chapter 2). There is a distinct difference between a definition that serves pragmatic intentions and the reality of organismic diversity, which fits only imperfectly into all recent definitions.

George Gaylord Simpson had already expressed this dilemma half a century ago: "Taxonomy is a science, but its application to classification involves a great deal of human contrivance and ingenuity, in short, of art. In this art there is leeway for personal taste, even foibles, but there are also canons that help to make some classifications better, more meaningful, more useful than others. . . ." (Simpson, 1961).

2
Why is there a Species Problem?

2.1
Objective of the Book

This book attempts to question the intuitive processes that are often employed in the classification of biological organisms. The author does not content himself with the diverse range of living beings being classified according to the principle of "order at all costs." After all, animals and plants are the products of evolutionary processes, which are governed by natural laws. Animals and plants are not postage stamps.

The problem with biological systematics is that evolution is not a uniform process; multiple selection processes of different natures play a role. Speciation depends on environmental conditions and on specific preferences in the choice of mating partner. Furthermore, speciation also depends on intrinsic properties of organisms' genomes that are unrelated to external conditions such as the environment or partner choice. Specific transposable elements in the genome that alter mutation rates can likewise be the cause of increased speciation rates (Prud'homme *et al.*, 2006).

Furthermore, not only are the conditions that cause speciation highly heterogeneous, but the processes referred to as speciation can also be manifold. Organisms can undergo alterations in traits as well as experience changes in mutual inter-individual connections, leading them to become new species. However, if a population exhibits changes in traits during the course of evolution, this is not the same process as when a population loses cohesion by dividing into two separate reproductive communities (see Chapter 6 and the anagenesis-cladogenesis conflict in Chapter 7). Individuals that are almost identical with regard to their traits might not be related to one another. Conversely, closely related individuals can have a markedly different appearance from one another (Chapter 5). Individuals that have markedly different appearances can mate with each other. Additionally, individuals that are almost identical in terms of traits can belong to completely separate reproductive communities (Chapter 6). How is it possible to unite such different processes into a common species concept? How is it possible to include the three taxonomically important classification principles (resemblance of traits, common descent and connectedness through mutual gene flow) in a consistent system? Can the existence

Do Species Exist? Principles of Taxonomic Classification, First Edition. Werner Kunz.

of multiple evolutionary processes be compatible with taxonomy's objective of constructing a consistent mode of classification?

Evolution proceeds with varying speeds. There are old and young species. If evolutionarily old species are widely spread geographically and their members are therefore distant from one another, then the organisms belonging to a species have evolved in different directions over time, regardless of the fact that they belong to the same species (Chapter 6). Geographically distant individuals within a species may differ substantially from one another both phenotypically and genetically (Garcia-Ramos and Kirkpatrick, 1997; Varga and Schmitt, 2008; Habel *et al.*, 2009). How can this be compatible with the use of genetic distance as a criterion for designating genetically different organisms as different species, which is a practice that forms the basis of the barcoding approach (Chapter 4)?

This book gets to the bottom of such questions. It explains and illustrates many biological examples that are associated with taxonomic problems. In doing so, special emphasis is placed on genetic foundations. However, an attempt is also made to philosophically substantiate theoretical conclusions. The question of how organisms are grouped both in our mind and in nature will be addressed. It is shown that a grouping according to intrinsic traits differs substantially from a grouping according to relational connections. It is also demonstrated that the formation of a class is completely different from grouping organisms as historically transient singularities, as individuals in the philosophical sense (Chapter 3). The philosophical term of the "natural kind" is dealt with, together with the question of "what is reality" in contrast to a purely mental concept. Special emphasis is repeatedly given to the question of "what is" as distinct from the operational principle of "what properties does something have" (see below).

In this book, species as objects of biology are often compared to the objects of chemistry, that is, atoms. In doing so, it is noted that the objects of biology are subject to evolution; they constantly change. In contrast, atoms are invariant. They would lose their class membership and thus become members of a new class if they changed in atomic number. In contrast, no living organism changes its class membership if it changes in some of its properties. A dipteran fly (the order of insects with two wings) remains a dipteran fly even if it experiences a mutation that causes it to develop four wings (the *Hox* mutation).

2.2
Can Species be Defined and Delimited from one Another?

The issue of biological species is one of the most curious problems of biology. Darwin once said something to the effect that everybody already seems to know what a species is: "No one definition has yet satisfied all naturalists, but every naturalist knows vaguely what he means when he speaks of a species" (Darwin, 1859). Everyone deals with this concept every day, but hardly anyone knows how to define the word "species." What underlies this discrepancy between spontaneous subjective certainty ("but that's all obvious!") and the fact that no one can indisputably say what a species is?

Many people devalue the species problem by saying that there are many other objects in our lives and in the natural world that also cannot be unambiguously defined. For example, no one can clearly say what a gene is (Paulsen and Nellen, 2008). The modern concept of a gene is no longer exclusively restricted to protein-coding units in the genome but encompasses a wide range of genetic and epigenetic variations, without scientists being able to find a "dividing line" in the genome where one gene ends and another begins.

With regard to differences in the content and biological functions of genome segments, which are all included within the common concept of a "gene," and the vagueness of species delimitation, there is actually a parallel between the species concept and the concept of the gene. This parallel is, however, weak. No one would be as inflamed by this argument if it concerned deciding whether a particular genome segment is a single gene or two separate genes. Almost every geneticist would answer that it is unimportant whether these two genome segments are considered to be two separate genes.

The situation is different for biological species. For example, a bird watcher would become excited if someone questioned whether two bird species might be a single species rather than separate species. That the African elephant is not a single species, but breaks up into two species is viewed as a matter of such importance that it was reported in the journal "Science" (Vogel, 2001). A similar report of the splitting of a gene, thus far considered to be a single entity, into two separate genes would not be considered to be of such merit. What is the explanation for this difference?

At first glance, everything seems obvious. The biodiversity of organisms seen in nature is not a uniform continuum: living organisms are arranged into groups. If a number of tits are seen in the gardens of a village in Europe, it is not a continuum of tits that is observed. Instead, some are recognized as Great Tits (*Parus major*), while others are Blue Tits (*P. caeruleus*). Additionally, the diversity of traits among different organisms is not distributed uniformly; there are peaks and troughs in the distribution of traits. It can immediately be seen that biodiversity falls within structured groups that can be referred to as "species."

The first hints of trouble that emerge from this apparent unambiguousness are the smooth transitions between species, that is, the blurry valleys among the peaks. There are high valleys with rugged slopes and flat mountains with gentle slopes. There might even be some hint of valleys between two peaks. There are no deep crevasses, provoking the question of how deep such gashes have to be to separate one mountain into two mountains. The question arises of whether certain stones at the bottom of a valley belong to the mountain to the left or to the right, simultaneously to both mountains, or to neither of them. Mixed zones are problematic with regard to group formation.

Yet it seems that the phenomenon of gradual transitions between mountains does not change the fact that there are mountains. Almost no one would infer that mountains do not exist based on the occurrence of gradual border zones. Nevertheless, a few taxonomists have drawn just that conclusion: owing to the overlap between two species and the resultant impossibility of assigning these hybrids unequivocally to one of the two species, they infer that species may not exist at all

(Mishler, 1999). However, indistinct borders do not seem to contradict the existence of species. If species did not exist, it would not even be possible to speak of the boundaries between them. Colors are not sharply delimited from each other; blue merges rather smoothly into green, but hardly anyone would draw the conclusion that colors do not exist.

2.3
What Makes Biological Species so Special?

The species as a taxonomic unit has been one of the most controversial topics in biology for over 150 years. As an article in EMBO Reports from 2000 states, "Modern biology has many triumphs to celebrate, but a generally applicable species definition is not among them" (Martin and Salamini, 2000). The authors go on to state, "It is perhaps biology's most grotesque concession that 140 years after the publication of Darwin's 'The Origin of Species' we still do not know exactly what those things are whose origin the theory of evolution explains."

There is hardly a biological topic that is as controversial as the concept of biological species (Mallet, 1995). Opinions on the subject of the species problem range from "this all is nothing new" to "this all doesn't help" or "I consider this matter purely pragmatic" right up to publications with titles such as "A radical solution to the species problem" (Ghiselin, 1974) and the observation that the quarrel about the correct species concept can attain dimensions similar to a religious war (Peters, 1998).

Why is the biological species so special? If the matter is considered *purely pragmatic*, why do two bird watchers argue passionately about whether a bird they have just seen belongs to one species rather than another? Two geographers would never argue with such passion about whether a mountain peak they are viewing belongs to one mountainous massif or to the adjacent mountain range.

This everyday observation already encompasses the assumption that humans approach species with preformed expectations (Atran, 1999). Humans appear to hold the firm belief that there are species in nature; just as we believe that the sun rises in the morning (see below). As early as half a millennium ago, Copernicus taught us that the sun does not rise, but we nevertheless perceive that it does, and we follow this perception even in our everyday language. Over 150 years ago, Darwin put forth the conviction that the biological species is a human construct, a position that is now presented again by many biologists and biophilosophers (Heywood, 1998), but we still believe unshakably that there are fixed species in nature that exist as natural entities.

More often than they do with other scientific problems, scientists dealing with the species problem encounter a complete lack of understanding rather than difficulties of comprehension or indifference (i.e., "what is this all about?"). The species problem is less a matter of complication than a matter of acceptance. On a daily basis, there are no significant problems that involve the concept of "species." Therefore, many people, even many biologists, do not understand that there is a "species problem," meaning that even experts do not agree what a species actually is. The issue is not how to distinguish two species from each other diagnostically but what a species is in reality (see below).

The species problem is sometimes viewed less as an epistemiological problem and more as a kind of paradox in the mode of the classical Greek paradoxes of the runner Achilles not being able to overtake a turtle and that you cannot "step in the same river twice." Instead of sounding out the underlying error in reasoning, the pursuit of this problem is often spontaneously classified as "nonsense."

In fact, some questions regarding the essence of species border on paradoxes. Three examples follow.

1) Species-less single organisms: If one accepts the species concept as a gene-flow community, then this kind of group formation is not a class formation (Chapter 3). Instead, it is a relational group formation of mutually reproducing organisms (Chapter 6). This unambiguously means that whether an organism in a community belongs to a species is defined only by the relationship of the organism with at least one other individual. Consequently, the paradoxical-sounding question arises: can the last survivor of a species on the verge of extinction actually belong to a species, as it belongs to no community of mutually reproducing organisms? This question sounds paradoxically, but it is not unjustified.
2) Origination of a new species without any alteration in traits: The cladistic species concept is based on the strict logic of monophyly. That is, every cladistic bifurcation of the original stem species into two daughter species means the end of the stem species and the origination of two new species that did not exist before the cladistic split (Chapter 7). Once one species splits into two separate groups, both groups have to be defined as new species, not only one of them. This clearly means that the original stem species never can survive a cladistic bifurcation. The logic of cladistics does not allow the survival of the stem species (being a kind of "main branch") after the lateral branching of a daughter species (being a kind of "side branch") because sister taxa must always be of the same rank. There are no main or side branches. That is, the original stem species cannot remain in existence if a lateral branch splits off, as this would contradict the principle of monophyly (Hennig, 1966) (Figure 2.1). From this logic, it follows that every bifurcation must be the end of the stem species.

   This results in a conclusion that is consistent, although it sounds paradoxical: if a species has a wide geographical range, such as throughout Eurasia from Spain to Korea, and a small population splits off and becomes reproductively isolated (daughter species No. 1) on a Korean island at the periphery of the geographical range of the species, then all that remains of the population throughout Eurasia simultaneously becomes a new species (daughter species No. 2), without having changed even minimally.
3) We can never know the phylogenetic age of a species. As a matter of principle, the age of a species cannot be determined. The rules of cladistics demand that the origin of a species is always based on the bifurcation of a stem species into two daughter species (Chapter 7). There is no other means of species formation, which leads to the conclusion that a species is always as old as the last bifurcation.

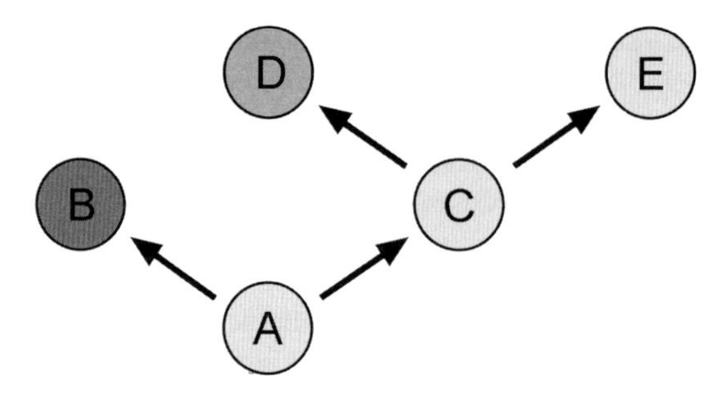

Figure 2.1 The cladistic species concept is based on the strict logic of monophyly. Every cladistic bifurcation of a stem species into two daughter species means the end of the stem species (*A*) and the origin of two new species (*B* and *C*). The stem species (right line, white circles) does not remain in existence, even if it does not undergo a change in its traits during the branching events from group *A* to *E*. *A*, *C* and *E* are different species. The apparent "side branches" *B* and *D* (shaded circles) cannot be considered as side branches, because the principle of cladistics implies that both daughter branches are of equal rank, contradicting the attempt to differentiate between main branch and side branch.

However, there are bifurcations that may not yet have been discovered because one daughter species immediately became extinct and only the second of the two daughter species continued to survive until the present. The surviving species is therefore given an incorrect (more ancient) age as long as the bifurcation of the extinct second daughter species has not been discovered. This is illustrated in Figure 2.2. Species *B* became extinct, which means that species *A* is younger than species *C*. As the possibility of an undiscovered lateral bifurcation can never be ruled out, the age of a species can never be determined.

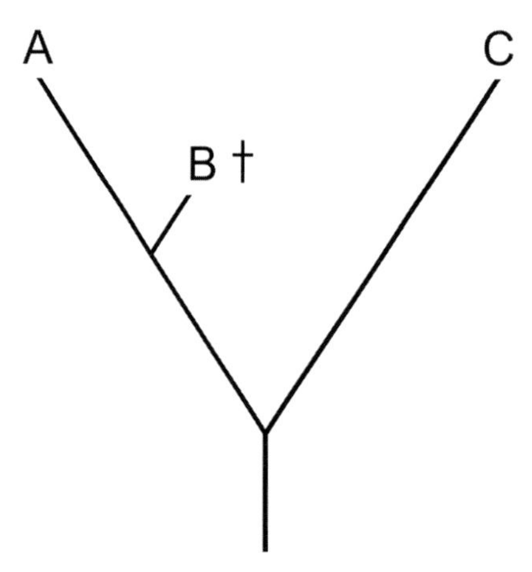

Figure 2.2 The law of cladistics implies that it is never possible to determine the age of a species. A species is always as old as the last bifurcation. It cannot be ruled out that the splitting of a side branch (*B*) has been overlooked because the organisms of that branch may have become extinct rapidly. However, this split means that species *A* is younger than species *C*.

These conclusions indeed sound paradoxical at first; however, this is only because they contradict spontaneous intuition. Intuition, though, is no reason to have an adverse attitude toward these conclusions. It is worthwhile to look for purely theoretical solutions in science (and not only in philosophy). Albert Einstein said something to the effect that he always desired to be able to explain empirical observations made in nature with a theory. This theory must above all satisfy the condition of being in accordance with laws of thought and not collide with them (cited in Fischer, 2005). Is there a theoretical basis for "morpho-species" (a species defined only by its morphological traits)? Can the theory of "morpho-species" be true or false? Can the theory of "morpho-species" be falsifiable in the sense of the philosopher Karl Popper? If not, is working with a "morpho-species" scientific?

Who can answer point-blank why the wolf and the fox are different species while the mastiff and the dachshund are not? The renowned rotifer expert Walter Koste (Germany) asked me several years ago by what right does he, cumbersomely and with technical effort, describe new rotifer species that are distinguished by only a tiny, hardly visible bristle, while he can distinguish some of his fellow humans comfortably by their hair color or their respective blood group. What is the answer to this?

How should the conflict between anagenetic classification (classification of species according to trait changes along the temporal axis) and cladogenetic classification (classification of species according only to bifurcation) be dealt with? During evolution, organisms undergo modification of traits in qualitative changes referred to as "anagenesis" (Chapter 7 and Figure 2.3). However, this is not the only type of change that occurs. Groups of organisms (species) also do something different from the modification of traits in the course of evolution: the phylogenetic lineage splits off into separate daughter branches. This is not a qualitative change but rather a numerical, quantitative change known as "cladogenesis." Qualitative and numerical changes are markedly different evolutionary processes and cannot be measured by the same yardstick.

Now we are confronted with a very difficult decision: is it anagenesis (the change in traits) or cladogenesis (the purely numerical type of change) that represents the origin of a new species? Or do both processes ultimately constitute the origin of new species? In the latter case, there would be two different kinds of speciation and, thus, two different kinds of species. A species cannot therefore be one and the same thing. Evolution pursues different modes of alteration (Mayr and Ashlock, 1991), which is why a conflict arises between anagenesis and cladogenesis. This conflict appears to be irresolvable, as it implicitly contains the premise that both trait change and bifurcation are simultaneously classified as speciation (Peters, 1998).

2.4 Species: To Exist, or not to Exist, that is the Question

There is no agreement about what a species is, and there is no agreement about whether species are artificial constructs of our mind or whether they really exist (Mishler, 1999). How then can someone claim to count species, to speak of reductions

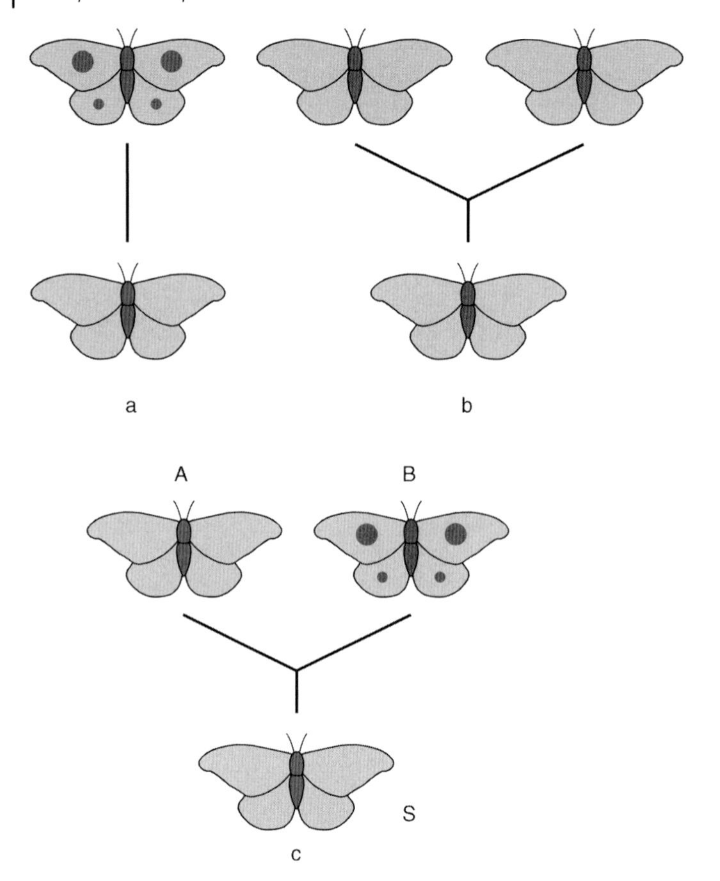

Figure 2.3 Evolution consists of two different processes. A species in a phylogenetic lineage may acquire new traits without bifurcation into daughter branches (anagenesis) (a), or it may bifurcate into two separate branches (cladogenesis) (b). Furthermore, the branching of a phylogenetic lineage may be accompanied by alterations in traits (c). *S* = stem species; *A* and *B* = daughter species.

in species numbers or to designate certain regions as rich in species and others as poor in species? These are questions that initially assure only one thing: evading the "species problem" or even declaring it unimportant or pointless represents a superficial attitude that cannot be justified. It is unavoidable that this problem must be faced as both a philosopher and a scientist. If speciation is to be any different from ordinary evolution, we must have a clear definition of species (Mallet, 1995).

Without a doubt, it is of practical value for taxonomists to attempt to keep House Sparrows (*Passer domesticus*) and Tree Sparrows (*Passer montanus*) separate, but is this classification more than a useful tool to distinguish something when you do not even know what it is that you are distinguishing? What are these groups that the taxonomist calls "House Sparrows" or "Tree Sparrows?" Evidently, there are single organisms, individual sparrows, and at least during certain seasons, groups of House

Sparrows and groups of Tree Sparrows, yet what is actually the taxonomic group that we refer to as a "species?" Does such a group even exist? Imagine that there were no intelligent humans - would all House Sparrows on Earth then constitute a group? Every human can group together single objects that are similar with respect to one property (e.g., being colored red), and most humans instinctively believe that such a group (the group of "reds") really does exist, but is this group not just a figment of our own imaginations (Chapter 3)?

Do species actually exist beyond the human imagination, or are we helpless victims of our mind's predetermined way of thinking, which instructs us from birth to think in terms of classes (Atran, 1999)? A child first sees a Volkswagen and learns that this is a car. Then he sees a Mercedes and is told once again that this is a car. He finally learns that a truck is a car as well and thus begins to abstract and learn to handle the concept of a car. In practice, he will not have any problems with this concept, but does the "car" exist in reality in an objective fashion beyond human classification and concept formation, even independently of our reasoning?

Many people would answer no and that the "car" as such is an abstract concept. Why then is the biological species concept at first glance clearly comparable with the car example, which is more likely believed to exist? Why else would someone mourn a species' extinction or speak of the species diversity of a certain region if the matter only involved organisms, not species? Is the case of "species extinction" definitely not the same as a reduction in the number of organisms? The answer is no; it involves species. Hardly anyone would agree that we are talking about self-constructed classifications here. Let us then double the number of species by applying stricter classification principles; species extinction has instantly been compensated for again.

The traits that are presently used to divide organisms into groups are not the complete set of traits that an organism possesses. Instead, these traits represent a restricted selection of all of the organism's traits. It is certainly possible to deviate from these traits and use other traits for classification. Different delimitation criteria will be applied using a markedly different choice of traits, and this will result in different species numbers.

For example, what principle dictates that we do not classify primates only by their blood group type? Some human beings would then belong to the same group as some rhesus monkeys, while other humans would be part of another group together with other rhesus monkeys because the groups are distinguished unambiguously by their blood groups from each other. Initially, this sounds absurd, yet who would immediately be able to explain the underlying theoretical principle of why "blood group taxonomy" of this kind is not "allowed?"

Imagining a simple genealogical tree (Figure 2.4), you can combine branches *B* and *C* in the same group because organisms belonging to *B* and organisms belonging to *C* share the same or similar traits. Many people would consider such a group artificial because the trait resemblance of *B* organisms and *C* organisms is based on artificial trait selection. As biological organisms differ with regard to thousands of traits, each arrangement according to trait resemblance inevitably has to be subjective because there is a possibility of forming markedly different arrangements by choosing markedly different traits.

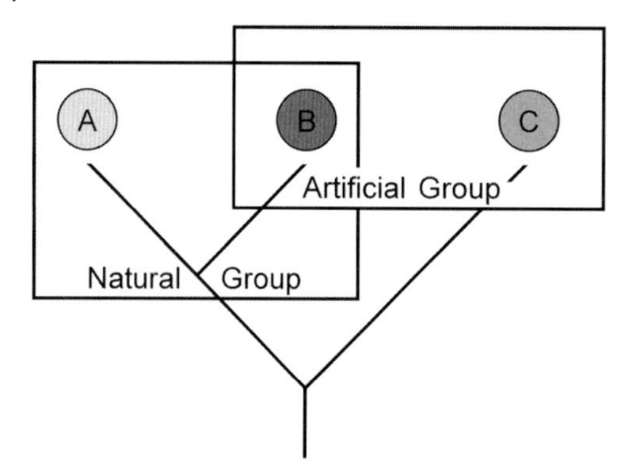

Figure 2.4 Taxonomic classification may be performed based on an assessment of trait similarity (*B* + *C*) or phylogenetic relatedness (*A* + *B*). The grouping of organisms based on trait similarity results in artificial groups (*B* + *C*), whereas the grouping of organisms by kinship results in natural groups (*A* + *B*). It may result in inconsistencies or contradictions to combine both grouping attempts into a common taxonomic system.

However, there is a different mode of arrangement. You can also combine branches *A* and *B* in a group not based on trait resemblance, but because they are more closely related (Figure 2.4). They are descendants of a recent common ancestor. Many people will consider such a group to be natural because common ancestry is not the product of a human sorting mechanism. There therefore appears to be natural groups, in contrast to artificial groups (Chapter 3).

Today, many biologists and natural philosophers share the opinion that a monistic species concept, that is, "the species," cannot exist (Wilson, 1999). The species is viewed pluralistically, meaning that it depends entirely on the respective concept of how classifications and demarcations are performed. This is frustrating and means that in principle, species cannot be counted (Hey, 2001). Every statement of species number depends on the formation of the respective concept, especially if the species numbers of different phyla of animals and plants are compared to each other (Mishler, 1999). If a certain family or order of rotifers contains 200 species, but a certain family or order of mammals contains only 30 species, then a comparison of these species numbers is futile, but it is made anyway.

This is even true for much more closely related groups. The species numbers of butterfly families and the species numbers of ant families are therefore not comparable. Butterflies are combined into species in much larger entities compared to ants. Butterfly taxonomists hold to subspecies and combine several subspecies into a common species in many cases, while ant taxonomists, following the advice of E. O. Wilson, award the rank of species to many subspecies (Jim Mallet, personal communication, 2008). A statement such as "800 ant species but only 200 butterfly species live in this tropical region" hence becomes meaningless. Imagine a physicist saying "How many kinds of atoms and how many elements there are on Earth completely depends on our classification principle."

If the species concept is, or even has to be, pluralistic, then species cannot exist in reality. This is because pluralism is an expression of different human viewpoints being evaluated differently, which leads to artificial species concepts in every case. The fact that several different species concepts are used concurrently even today (Mayden, 1997) means that only one of the concepts can be true, or that there are no species in nature, or that the word "species" represents several things simultaneously. The last case would mean that a species does not have a distinct ontological structure because its existence traces back to different biological processes (Ereshefsky, 1999).

Many taxonomists and biophilosophers are currently of the opinion that species do not actually exist in nature (summarized in Wilson, 1999), whereas others consider species to be real (e.g., Mayr, 2000 4643; Ghiselin, 1997). Dissension reigns. Whether taxa such as *Homo sapiens* or *Drosophila melanogaster* exist in reality is certainly a justifiable question.

2.5
The Reality of Species: Ernst Mayr vs. Charles Darwin

Those who consider that species do not exist reapproach the position that Darwin had assumed more than a century and a half ago. Darwin considered that species were not real, representing a figment that we had created solely for the sake of our own convenience. However, the founders of the "synthetic theory of evolution" in the 1930s and early 1940s (also referred to as the "modern synthesis"), including Theodosius Dobzhansky, Julian Huxley, Ernst Mayr, Bernhard Rensch and George Gaylord Simpson, considered species to exist in reality (Mayr, 1982).

Darwin (1859) wrote in "On the Origin of Species" in 1859 "Hence, in determining whether a form should be ranked as a species or a variety, the opinion of naturalists having sound judgment and wide experience seems the only guide to follow." Darwin goes on to say "... I look at the term species as one arbitrarily given, for the sake of convenience, to a set of individuals closely resembling each other, and that it does not essentially differ from the term variety, which is given to less distinct and more fluctuating forms. The term 'variety,' again, in comparison with mere individual differences, is also applied arbitrarily, for convenience's sake. ... Varieties have the same general traits as species, for they cannot be distinguished from species ... There is no real or natural line of difference between species and permanent or discernible variety ... nor do there exist any features on which reliance can be placed to pronounce whether two plants are distinguishable as species or varieties" (cited in Mayr, 1982).

Darwin therefore treated species as the higher taxa (genera, families, orders and classes) that, according to current general consensus, are only artificial groupings constructed for the sake of convenience. In a letter to Hooker Darwin writes 1856, "It is really laughable to see what different ideas are prominent in various naturalists' minds, when they speak of 'species'; in some, resemblance is everything and descent of little weight - in some, resemblance seems to go for nothing, and creation the reigning idea - in some, sterility an unfailing test, with others it is not worth a farthing. It all comes, I believe, from trying to define the indefinable" (cited in Mayr, 1982).

Darwin's principal work bears the title "On the Origin of Species" (1859), which is misleading because, as noted above, Darwin did not actually believe in the existence of species. This choice of words is as unfortunate as if Copernicus had titled his attack on the geocentric world view "On the Origin of Sunrise" (Atran, 1999).

Ernst Mayr, on the other hand, who received the title "Darwin of the twentieth century" several times (Glaubrecht, 2006) (a title that Mayr certainly would have greatly enjoyed), opposed Darwin vehemently, sometimes almost polemically, by accusing him of having carried out a "complete turnaround in [his] species concept" (Mayr, 1982) between the 1830s and the 1850s. However, today it is precisely this "turnaround" that is viewed as a probable step in the right direction. Darwin's gradually growing realization that the species as a reality does not generally withstand critical and consistent considerations bears witness to his foresight.

Mayr views Darwin's renunciation of the species as a reproductive community and his orientation toward the conviction that there are no species as a relapse triggered by Darwin becoming increasingly influenced by the botanical literature and correspondence with his botanist friends during the course of his life. Darwin himself says that "All my notions about how species change are derived from a long continued study of the works of (and converse with) agriculturists and horticulturalists" (cited in Mayr, 1982). While Mayr sees this as something akin to defection, the tables definitely need to be turned here. Darwin's foresight resulted from his ability to include all organisms in his theory and, especially, to factor in plants. Conversely, Mayr could be accused of having derived his species concept mainly from his knowledge of birds and thus having neglected many other groups of organisms as well as, and above all, the plants.

Among botanists in particular, extreme attitudes related to denying any existence of biological species prevail. Brent Mishler, director of the herbarium in Berkeley, titled a comprehensive article in 1999 "Getting rid of species?" (Mishler, 1999), and Konrad Bachmann (formerly at the Leibniz-Institute of Plant Genetics and Crop Plant Research in Gatersleben, Germany) dubbed a publication in 1998 "Species as units of diversity: an outdated concept" (Bachmann, 1998). The biological species is referred to as worn out, hopelessly vague or even evidently wrong when compared to the persistent idea of ether in space.

2.6
The Constant Change in Evolution and the Quest of Taxonomy for Fixed Classes: can these be Compatible?

When I published an article several years ago under the heading "Was ist eine Art?" (What is a species?) with the subtitle "In der Praxis bewährt, aber unscharf definiert" (Approved in practice, but vaguely defined) (Kunz, 2002), I received a call from a computer scientist alerting me to the notion that there are no such things as vague definitions. Definitions are not discovered but produced. Definitions are not subject to evolution and are thus precise to begin with. Only objects (not definitions) can be vague and can therefore fit well, badly or not at all into a given definition. However, the definition itself is always precise. This is also true for taxonomical groupings of organisms into classes. The classes are precise, but the organisms fit more or less precisely into such classes (Chapter 3). From this reasoning, it follows that there are organisms that do not fit into a class at all. These are species-less organisms.

In the 1930s, Sewell Wright (1935) created an evolutionary metaphor that has become known as the "adaptive mountain range" or "adaptive landscape". This symbolic landscape consists of numerous adaptive peaks separated (but only incompletely separated) by valleys. Biological species are considered to constitute such peaks. An "adaptive peak" represents a state of equilibrium in which the gene pool exhibits an allelic frequency distribution that has become stable because of optimal success of a population in the environment. Even within a stable environment, several adaptive peaks are possible for a given species (which is known as intraspecific polymorphism in taxonomy; see Chapter 5), though natural selection generally tends to hold a species to a single peak. This model of biological species remains currently applicable and worthy of discussion (Mallet, 2010).

The main problem with this model is its smooth boundaries. The opinion exists that vague boundaries between two objects are at odds with these objects really existing. If there is overlap, then there are objects in the intersection area that cannot be assigned to either group. Accordingly, the conclusion is drawn that the two overlapping object groups are indefinable and therefore do not exist. However, this position must be opposed by stating that there is a difference between the definition of a group and the group as it really exists. The definition is an artificial template that is imposed on the reality of the defined group.

A definition, like a class (Chapter 3), is always fixed, sharply delimited and does not overlap with the definition of another object. Classes are not subject to evolution. However, the object, meaning the group itself, or the species in this case, is not the same thing as the definition of the group. The group itself can be vaguely delimited and overlap with neighboring groups. The only consequence is that single objects belonging to the group in the intersection area cannot be assigned. In spite of this, groups themselves can very well exist. Colors are a good example of entities that overlap fluently. There are no sharp demarcations between colors; nevertheless, they do exist. This again makes it clear that there are organisms that cannot belong to a defined species. They are either species-less or they belong to two different species simultaneously.

It is therefore important to carefully distinguish whether a grouping is a group existing in reality or a theoretical concept created by humans for the sake of our own need to classify things. How do you recognize a group existing in reality, and when is a group a theoretical construct? The distinction is not easy to make and continues to be debated. The problem of the existence of groups dates back to Plato and was the focus of fierce disputes regarding the problem of universals centuries ago.

The objects of taxonomy, individual animals and plants, are subject to constant evolutionary change; that is, they are vague and variable. All individual House

Sparrows (*Passer domesticus*) existing on Earth, including their ancestors and descendants are, or were, subject to evolution. They mutate and are selected. What, then, is the House Sparrow taxon? Taxa themselves cannot be vague or variable. Classes in an Aristotelian sense are fixed concepts and cannot therefore be vague or subject to evolution (Mahner and Bunge, 1997). The class concept "species House Sparrow" cannot mutate and be subjected to alteration; if this did take place, the concept would become a new class. The taxon as a class is to be understood as a human-constructed template that is temporarily imposed on the reality of organismic diversity. There will always be non-assignable organisms. The vagueness lies in the fact that the assignment of existing organisms to a predefined class is precarious, or even impossible in boundary cases. However, the vagueness of assigning particular objects to a class does not lie in the fact that the taxonomic class itself is vague.

A taxonomic class is always a theoretical construct. In reality, there are no traits that, as a necessary and sufficient criterion, bind particular organisms together in a group while excluding others. Of course, you can find a particular identification trait that distinguishes a House Sparrow (*Passer domesticus*) from a Field Sparrow (*Passer montanus*). Where is the problem in that? All Field Sparrows have a black spot on the cheek that House Sparrows do not have, so they look different. But do all Field Sparrows really have that spot or is it only the vast majority of them?

In attempting to find a set of organisms that belong to a single species due to being characterized by a particular essential trait, you will always encounter some organisms that do not exhibit this trait but still belong to this species. This is a far-reaching declarative statement, as it brings the fact that species are not defined by traits to light. Trait equivalence cannot be an essential requirement for species membership. The possession of a trait can never be understood in a way that means that every organism that presents this trait must absolutely belong to this and no other species, or conversely, that every organism with differences with regard to this trait must absolutely belong to a different species. In other words, species are not "natural kinds" in the philosophical sense (see Chapter 3). If you attempted to define a species by means of essential traits, you would have to exclude any organism that displays a different trait from the species just for this reason, yet this is not done.

The creation of a class is an attempt to match objects of empirical knowledge with an explanatory template. This template is human constructed, although several criteria used as components of the template are taken from evolutionary or other natural characteristics. Taxa in the sense of Linnaeus cannot be subject to evolution. The idea of a biological class in the Aristotelian and Linnaean sense as existing in reality and the reality of organisms being subject to evolution, in a state of constant change with regard to their traits, are incompatible with each other. If taxonomy and evolution must be united, this can only be done through the concept of the human-constructed class. The fixed type in the Platonic sense, which Linnaeus realized in nature and described in his taxonomy, does not exist and was abandoned by Darwin.

Taxonomists prefer to set aside these and similar implications (Neumann, 2009). They often act as if they are describing a virtual world with respect to taxa that exists independent from our mind in the outside world. However, although the Darwinian revolution has markedly revolutionized taxonomy (Dupré, 1999), this discipline still seems to cling to the traditional Linnaean concept of an essentialist philosophy.

Darwin himself noted this and ultimately denied the existence of species: "Since species continue to evolve, they cannot be defined, they are purely arbitrary designations. The taxonomist no longer will have to worry what a species is. . . . This, I feel sure, and I speak after experience, will be of no slight relief." (cited in Mayr, 1982). Darwin thus realized that he had entered into a contradiction between the purpose of classification and the theory of evolution he had founded.

Mayr criticizes this attitude of Darwin's by making the important point that Darwin's perspective only refers to the typological species concept. It is just the perception of a species as a class that has been turned into a human-constructed artificial product by the theory of evolution, not the perception of a species as a group of organisms held together by cohesive relationships. If a species is considered as a group with a common genealogical descent, that means in which there is coherence between parents and their children or if a species is considered as a community where there is gene flow via sexual partners, the species does not conflict with the theory of evolution. Such cohesive groups unequivocally prove themselves to be natural and not human constructs. Descent coherence and sexual coherence unquestionably specify real groups. However, if you do not view the species as a class that is sorted on the basis of trait resemblance, it must be admitted that trait resemblances, although they might play an important diagnostic role, are never the ontological criterion for the status of a species (Sterelny and Griffiths, 1999) (see the Section below "It is one thing to identify a species, but another to define what a species is").

2.7
Can a Scientist Work with a Species Without Knowing what a Species is?

Can a scientist work with species without knowing what species are? This question is not easily answered. Initially, everyone would say "Of course not." However, it may be possible for a taxonomist to sort and group species. This would then be a purely pragmatic approach, just as a doctor can cure a disease without being able to define what a disease really is. Thus, it is apparently possible to use things successfully without knowing what these things are. It is indeed difficult to say what a disease is. What trait distinguishes disease from health? Try to define both what disease is and what health is. Is a pain in the calf a disease? Is age a disease?

Francis Crick, codiscoverer of the molecular structure of DNA, once said something to the effect that you do not win battles by discussing what a battle is. In other words, you can win a battle without knowing what a battle is. In contrast, the biophilosopher David Hull said that "We must resist at all costs the tendency to superimpose a false simplicity on the exterior of science to hide incompletely formulated theoretical foundations" (Hull, 1970). Many philosophers of science warn against a scientific mode of reasoning that is defined completely operationally without being founded on a theoretical basis (Hull, 1968).

It is very peculiar that a concept such as that of the species has been successfully in use for centuries, even though it is not consistently defined. Different people, even scientists, understand different things by the term "species." A mammalian species is something different than a rotifer species or a bacterial species, not on the basis of content but on the basis of concept. Can the concept of "species" ever be defined consistently?

Most people understand a species as a "type," a "form" or a "kind." This can be seen in questions such as "what kind of bird is this" or "what type of butterfly is this, then" or phrases such as "two forms of crucifers grow in our garden" (Schilthuizen, 2001). Darwin noted that there is something peculiar about the species concept: it is used every day, though no one can define what a species is (Darwin, 1859).

This is a highly alarming matter for a scientist. That is, scientists work with species, and through this they achieve myriad scientifically reproducible results, as evidenced by thousands and thousands of publications. However, at the same time, they do not know what species are. Can a scientist, who is pledged to the truth, acquire insight into an object without knowing what the object is? This must ring a warning bell. The door seems to be wide open for hunches and instincts, which is dangerous. Alternatively, is it normal that scientific progress can be achieved using certain objects without knowing what these objects are?

2.8
The Species as an Intuitive Concept and a Cognitive Preset in the Human Mind

A (European) Blackbird (*Turdus merula*) is a bird that has a yellow beak and black plumage. However, there are also albinos or semi-albinos that exhibit either completely white or partially white plumage. They are still Blackbirds, with no taxonomist having any doubt about the classification. Interestingly, even laypersons and experts who have not thought about this problem before do not have any doubt that the albino Blackbird is a Blackbird. What, then, is a Blackbird? Why do humans not doubt that even a white Blackbird is a Blackbird?

Humans have held species concepts for as long as we have existed, whether we are aware of them or not. We classify the nature surrounding us into discrete groups on a purely intuitive basis. Grouping of organisms was initially merely a matter of survival. Humans needed to know what predator we had to be wary of and what berries and fungi we could eat. We use species concepts in everyday life to predict the properties of organisms and react accordingly. Species concepts are thus everything but intellectual games of eccentric scientists (Neumann, 2009).

That the various animals and plants in our environment do not constitute an uninterrupted continuum but belong to distinct groups seems to be a component of innate human knowledge. Once good or bad experiences have been associated with an organism, humans are spared going through the same experience a second time because the conclusion regarding the good or bad properties of the organism is automatically transferred to a second organism with a similar appearance. We are all genetically programmed taxonomists. This realization might sound encouraging to the taxonomist, but it is highly dangerous because our spontaneous sensations are cognitively programmed. They are thus able to deceive us in the same way as the idea of a flat Earth or sunrise.

The physicist knows better than the biologist with respect to space and time that we are easily misled by the appearance that nature presents directly to our senses and that reality is often different from what we perceive. The intuitive knowledge that two events that humans on Earth observe at the same time in the universe also happen at the same time for an observer who stands at a different point in the universe is a useful subjective form of organization by our mind. Since Einstein, however, we have known that this is not actually the case. The concept of contemporaneity is not an objective reality that exists independently from space. We should be vigilant that the classification of species does not turn out to be a similar situation in which we are fooled by spontaneous perception. In any case, caution is advised with regard to trusting intuition.

The cognitive scientist Scott Atran (1999) explored the question of the extent to which the feeling that certain organisms belong to groups and our certainty that there is a natural relatedness underlying this have a predetermined genetic basis. Atran arrived at the conclusion that we are all born typologists, whether we want to be or not.

In years of exploration, Atran compared native Mayans in Guatemala with urban students of the US Midwest. The first group did not have formal education but constantly encountered animals and plants in nature, whereas the second group had good formal education but almost no contact with nature. Humans of both cultures classify animals as well as plants into groups. For example, several thrush species and oak species are combined into natural groups because of their resemblance without these categories ever having been learned. It is automatically assumed that there is a sameness that holds the organisms together in a “natural” unit. In contrast, inanimate objects, such as stones, are not perceived in a way indicating that natural cohesive groups are present. Remarkably, the Itzaj Maya from the middle-American rain forest are no different in this respect from city residents in Michigan.

Classification of organisms into taxa does not seem only to be a product of education but, rather, seems to be inherent to human beings. Across a number of cultures and educational systems, there seems to exist a mental representation of the biological world in our mind independent from any actual experience that is comparable to our cognitive preset with respect to the concept of space and time. No human who has an awareness of the concept of time ever had to learn what time is.

Ideally, selection would have led to humans being better able to deal with the organisms that they encounter in their environment. We can almost instantaneously identify organisms as enemies to be dealt with (without having to be individually subjected to the harmful experience), or identify them quickly as harmless partners with which we can live, or as exploitable food sources (without having to check in every case whether a fruit tastes bitter or is poisonous). It is essential for the survival of a newly hatched chicken, which is easy prey for a raptor, that it can distinguish an Accipiter raptor (Goshawk or Sparrow Hawk) from a harmless dove, without the chicken ever having had an opportunity to learn this. In this process, the type of dove or hawk flying over the chicken is unimportant. What matters is the group identification.

A strong tendency to perceive different organisms as groups with a similar nature has therefore been proven to be advantageous for selection and has prevailed because of this. This is what Atran refers to as the "generic species" concept, that is, the "automatic" identification of biological group membership at the genus level (Atran, 1999). A cognitive structure independent from individual educational culture inherently specifies the scope and the restrictions related to categorizing animals and plants such that every organism must belong to one and only one essential group. Apparently, we cannot think in any other way. This has resulted not only in individual groups of organisms being given names, but also in different groups of organisms becoming associated with fixed properties (Heywood, 1998).

This is also what is understood by the concept of "folkbiology" (see below). Our mental preset related to how groups of organisms are perceived leads to especially contentious behavior. Compare a geographer with a taxonomist. The former argues with a colleague about whether a particular small elevation between the mountains Jungfrau and Eiger in the Swiss Alps is a new, "distinct," "real" mountain. The taxonomist analogously argues with a colleague about whether a newly discovered population is a new, "distinct," "real" species. The dispute among the taxonomists will undoubtedly be more passionate and vigorous. Taxonomists are much more excitable about whether a group of organisms could represent a new species than geographers are about whether an additional elevation indicates a distinct mountain or not. Why is there much more excitement and argument in the case of species delimitation than in the case of the delimitation of geographical entities?

Most taxonomists admit that there are no sharp boundaries between species; they blend into each other more or less smoothly. What then is the difference between the individual mountain tops within a mountain range and the individual species of a taxonomic family? Why does the question of how many species the family of *Felidae* (cat-like animals) consists of and whether it is more than the family of *Canidae* (dog-like animals) appear to be serious and meaningful, while the question of how many mountains the Black Forest in Central Europe encompasses and whether it is more than are present among the nearby Vosges Mountains appears to be inconsequential?

The extent to which we are subject to the compulsions of classification based on perceptible traits might also explain the remarkable fact that Mayr's species concept of the reproductive community and all that it entails has not yet established itself, even though it has been taught in schools for over half a century as the actual biological species concept. The species concept as a reproductive community defines species membership unequivocally as a relational connection (Chapter 3). This means that an organism can belong to a species only if it has a connection to a second organism, just as a human can only be a brother if he has siblings, and a human can only be a neighbor if somebody lives next to him. The logic of this relational connection implies that group membership does not primarily involve traits. Nobody claims that the hermit has to alter his traits to become a neighbor. Traits can never define what a species is; they can only help us to identify a species provided that it is already known by other criteria that it is a species.

Contrary to this, it is always automatically assumed that the emergence of a new species must involve changes in traits. Although Mayr's species concept has been accepted as commonplace by biologists for half a century, it has never fully established itself in our consciousness. The need to associate specific traits with species membership has prevailed. To understand the origin of a new species as an alteration of the relational connections between organisms, without primarily being associated with the alteration of traits, is difficult for every human (Chapter 7).

2.9 Taxonomy's Status as a "Soft" or "Hard Science"

With the exception of neurobiology, taxonomy is the one field of biology that as a science, requires a bridge to philosophy (Mayr, 1982). This is because if taxonomy aims to achieve more than just the collection and classification of organisms, philosophical questions, such as what a group is, what a class is, or under what conditions a group can be considered to exist in reality in contrast to being just a fictional concept, are involved.

The species problem is a scientific problem that lies at the intersection of scientific and philosophical research interests. Unfortunately, the two disciplines have little contact and speak different languages. The common result arises that what is not understood is rejected. This hinders scientific progress because both disciplines could benefit from each other.

Scientists are interested more in empirical data, whereas philosophers are more concerned with the conclusiveness of thought processes. However, the biological species cannot be explained through empirical data alone, unless there is no concern with knowing what a species is. It is precisely here that the crux of the matter lies. The philosopher always wants to know what a species is, while the scientist evades this question because he can apparently work with species without knowing what a species is (see above).

The large majority of publications dealing with species are either scientifically oriented or philosophical. Both perspectives can rarely be found together in a review article or book, yet empirically provided data should necessarily be supplemented with philosophical aspects. The taxonomist cannot be indifferent about whether species exist in nature as real entities or whether they are just constructs of our minds. This is because the species is a group in any case, and to form groups, it is useful to understand the logic and mental laws upon which group formation is based. These criteria represent the first step toward philosophy.

Objects with identical traits being sorted into a group or the members of a group being held together by mutual relational connections (Figure 2.5) are markedly different modes of group formation. It is of fundamental importance to be clear about whether species are groups of organisms held together by cohesive forces or are abstract categories (i.e., class formations existing independent of space and time; see Chapter 3).

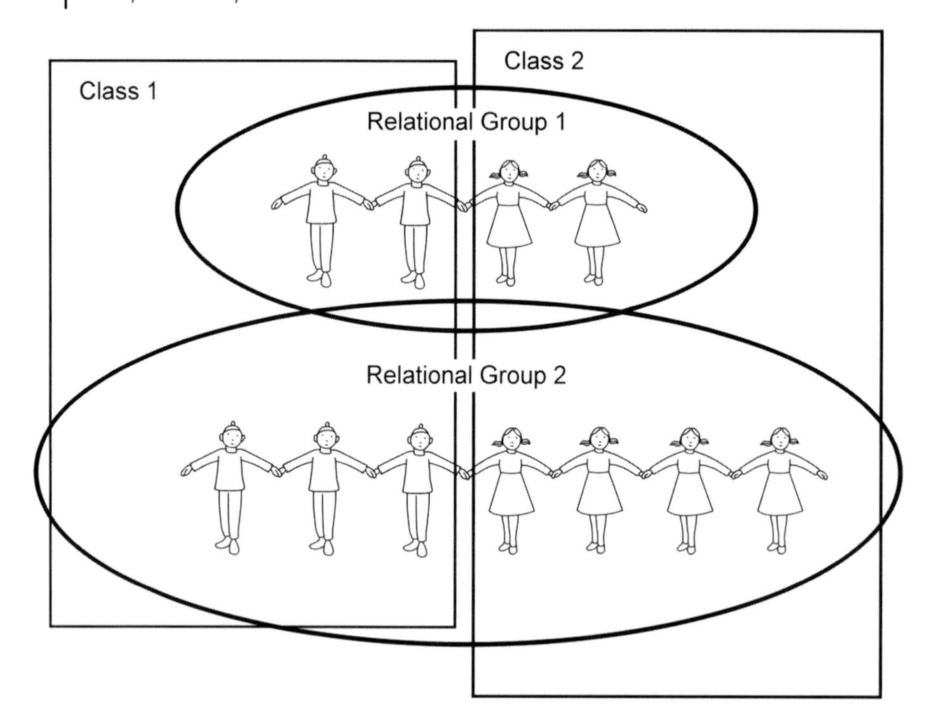

Figure 2.5 If a number of children are classified into groups, there are two alternatives. They may be combined based on trait equivalence (e.g., the two groups are boys and girls), or they may be combined into two groups that are linked by hand holding with each other. The first kind of grouping is a class formation resulting in an artificial group that is made by human classification efforts. Although boys and girls exist in reality, such groups do not exist in nature outside of human mind. In contrast, the second kind of grouping is relational grouping according to mutual cohesion between the individual objects. Such groups are considered to be realities.

Interestingly, many philosophers are bothered by the contrariness of the species concept. The paradox is that biologists are affected by the species problem but often ignore it, whereas philosophers are aware of the problem but are not affected by its consequences. Many biologists therefore retreat to the position that the species problem is philosophical and does not affect them; their aim is to identify a species diagnostically without defining it ontologically. However, diagnosis is not a species concept but is instead a recipe for the identification of different species (Sterelny and Griffiths, 1999) (see the Section below "It is one thing to identify a species, but another to define what a species is"). What is the use of this? It tells us more about human perception than about something existing in nature. If a species actually exists in nature, it should not matter whether humans can identify or distinguish it.

Scientists often become impatient and consider the philosophical dispute about the species problem as something that does not lead very far (Dieckmann *et al.*, 2005). Coyne and Orr request more empirical data on speciation and fewer debates about what a species is (Coyne and Orr, 2004). Pigliucci, in contrast, states that "philosophers of science could be extremely useful to the practical scientist, if only we would stop a moment to listen to what they are saying" (Pigliucci, 2002).

The downside of this controversial situation is no more and no less than the endangerment of the status of taxonomy as a science. If biological species are only a matter of diagnosis and sorting efforts, they cannot be subject to natural laws. This is reasonable grounds for suspecting that taxonomy is a science set apart from the other natural sciences, which have their own laws of thought (Griffiths, 1999). That biology is subject to different laws of thought than the rest of the natural sciences is something that Ernst Mayr also noted and often verbalized in his own way (Mayr, 1982). However, he saw this as something positive, as though this would speak for biology; he did not see anything alarming in this situation that would possibly endanger biology's, or at least organismic biology's status among the natural sciences. Although physics today is something markedly different than physics at the time of Newton (indeed, the discipline is so severely different that Newton's physics must be considered to be wrong to a certain extent) (Smolin, 1997), which is not true for biology, contemporary physicists view earlier physics respectfully and continue to teach it. In contrast, modern biologists, who are very much focused on the functioning of molecules, devalue organismic biology in a somewhat derogatory manner as "classical" and no longer teach it. This is a telling difference.

Atran (1999) and Bachmann (1998) refer to taxonomy as an intuitive "folkbiology." Biological organisms are intuitively combined into groups on the basis of their traits (Heywood, 1998). However, scientists must be careful not to become the victim of mental preconceptions (see above). Admittedly, evolutionary epistemology (Lorenz, 1977) teaches us that cognitive preconceptions are an adaptation to the requirements of optimally finding our way in the environment. However, a world view that is optimally suited to finding our way, orienting ourselves and working progressively is not the same as the truth. To find our way on Earth in everyday life, we do not need to know that the Earth is a sphere; the intuitive feeling of the Earth being a disk suffices. You have to carefully distinguish between "common sense" and provable facts. The mental representation of taxon groupings, which is not based on experience, may be an illusion (Atran, 1999).

In 1928, when Ernst Mayr studied the bird species living in the Arfak mountains in New Guinea, he charged the incumbent natives, who were experienced bird hunters, with shooting birds and bringing them to him. Every time a new bird was brought to him, he asked the hunter for the bird's name in the native local language. To his surprise, he realized that his intended work there had already been done. The natives distinguished, as did he, approximately 135 bird species (Schilthuizen, 2001).

This experience led Mayr to the conviction that species must be distinct entities in nature because people with different educational backgrounds and cultures classified birds using the same traits (Mayr and Ashlock, 1991). However, the opposite conclusion should actually be drawn. This astonishing perceptional identity speaks more to the similarity of the mental grouping and classification structures existing in the human brain than to 135 bird species actually existing in nature. The conceptions of laypeople in the field of taxonomy do not justify concluding that this mode of thinking is true; with exactly the same justification, you could defend the truth of Newton's physics against Einstein's theory of relativity and Planck's quantum physics, as Newton's doctrine is more intuitive and thus more acceptable for the layperson.

Taxonomy, the science of the systematic grouping of living organisms, is clearly believed by many people to be a discipline of "folk science." Sciences that consist to a substantial degree of conceptual structures and research strategies that are prescientific, that is, close to everyday life, can be referred to as "folk sciences" (Menting, 2002). They often seem appealing and esthetically pleasing to an extra-scientific lay clientele for precisely this reason. Not for nothing is taxonomy said to be a "scientia amabilis." The chances and dangers associated with such a discipline may not be ignored.

There are certainly reasons why taxonomy is not recognized as an equal-ranking science alongside other scientific disciplines, a status that contemporary taxonomy has to struggle for more than other sciences. Taxonomists feel that their discipline is not sufficiently esteemed. For an example of this opinion, see the biannually published newsletter of the German Gesellschaft für Biologische Systematik GfBS (Society for Biological Systematics). There is hardly another field in biology that is so disputed with regard not only to its findings and aims, but also to its right to exist; hardly any other field in the biological sciences is held in such low regard.

Taxonomy clearly has difficulties in being recognized as a "strict" science, and there are certainly reasons for this dilemma. It is not that there is a conspiracy of the mighty against the weak, that is, of the generous third-party-funded sciences against the inadequately financed sciences. There are also intrinsic reasons for this that are rooted in the discipline of taxonomy itself. This may represent a conflict between a "soft science" and the "hard sciences" (Pigliucci, 2002). Thus, it is not only the technical content of taxonomy, but most of all its epistemological foundations that are of importance.

2.10
The Impact of the Species Concept on Nature Conservation and the Allocation of Tax Money

The protection of species is an important task undertaken by nature conservation associations, but what species should be protected if there is no consistent opinion regarding what a species actually is? Should increased protection efforts and larger financial sums only then be employed if "real" species are endangered with respect to their existence, or should equal-ranking efforts be undertaken if they "only" involve the conservation of geographical races? This question is contentious (Moritz, 1994; Crandall *et al.*, 2000; Allendorf and Luikart, 2006).

The "Red Lists" of threatened animal and plant species generally only refer to species. Races are left high and dry, and it is considerably more difficult to receive tax money for the preservation of races than for the preservation of species. For example, the newly discovered Grey-Shanked Douc Langur (*Pygathrix cinereus*) in Vietnam first had to be assigned species status before funds for its preservation were provided (Jörg Adler, director of the zoo in Münster, Germany, personal communication). The status of a geographical race would not have sufficed. Typically, the sought-after label of a "real" species is awarded on the basis of DNA analysis. This is questionable because it is controversial whether this is a reasonable and sufficient criterion to determine the status of a species (Chapter 4).

The Red Wolf (*Canis rufus*), which is extremely endangered regarding its continued existence, is a rare inhabitant of North America. The Red Wolf was originally widely distributed in the southeast United States, but only a few remain today in Texas and Louisiana. After the American "Endangered Species Act," which is a law addressing species protection, was brought into effect in 1973, a debate arose about whether the Red Wolf was eligible for protection measures (Roy *et al.*, 1994). Because of its state of endangerment, there should have been no debate whatsoever regarding granting the Red Wolf the highest priority for protection.

However, DNA analyses showed that Red Wolf DNA contains the genetic material of both wolves (*Canis lupus*) and coyotes (*Canis latrans*). This resulted in a disagreement about whether the Red Wolf is actually an autonomous species. It was discussed whether the Red Wolf was instead a hybrid between wolves and coyotes (Chapter 6). If this suspicion were confirmed, would this justify canceling the efforts to protect and propagate the Red Wolf? Although it is very rare among animals, it has to be considered that a hybrid between two species can, in principle, be the origin of a new, third species (Chapter 6).

These examples sound a bit absurd. Environmental protectionists would be well advised to more strongly embrace the idea that the protection of biological diversity is different from the "protection of species" (Moritz, 1994 6058; Crandall *et al.*, 2000; Allendorf and Luikart, 2006). The Red Lists should be more than species lists. Multiplicity in nature has little to do with the species concept. Each of the present Red Lists is a document for the fact that intraspecific variability is carelessly neglected (Chapter 5).

The ethical value of preservation of biodiversity is not decreased by it becoming apparent that protective measures to benefit an endangered population are "only" benefiting an endangered subspecies instead of a true species. This realization has begun to prevail, such as in the fight for the preservation of rare races of domestic animals as well as endangered crop varieties. Here, the preservation of biological plurality has long since been concerned with a variety of forms and not with species.

2.11 Sociological Consequences of a Misunderstood Concept of Race

The idea that species are not only a construct of the human mind is justified by some taxonomists based on the fact that species are astonishingly applicable to reality. The realization that we are victims of our mind is not synonymous with the conclusion that a trait-oriented perception of species does not have anything to do with reality. As a biologist and Darwinist, one must of course ask the question of how the mental representation of the essence of a species enters our head. This likely represents a case of an adaptation of our mind to the requirements of everyday life.

However, it is not acceptable to reach the conclusion that this also reflects the objective state of the real world. Selective fitness-related advantages in the evolution of our brain have emerged through food competition or due to conferring combat-related advantages over rivals, but this is not tantamount to perceiving nature as it really is. A strong tendency to perceive varying individual organisms in the form of essentially distinct groups understandably offers selective advantages. Thus, fruits, prey and hostile animals or groups of humans could be perceived more quickly.

Human history, however, shows the fatal consequences of projecting a biological essence onto social groups or alien races. Human races are traditionally seen as genetically consistent but distinct from one another. This is an intuitive prejudice, and it is wrong (Sesardic, 2010).

The overemphasis on group differences as opposed to the often much greater variations among the individuals within a group ("what is part of another group has to be different") has led to race discrimination and ethnic "cleansing" based on instinct rather than reality. The disastrous history of Nazi Germany demonstrates how fatal the consequences can be if the concept of race is not understood properly. Even some leading scientists in human genetics misunderstood the race concept markedly at this time (Benzenhöfer, 2010), for example Otmar Freiherr von Verschuer, a leading professor in human genetics from 1936 to 1965 in Germany (Müller-Hill, 1999).

In contrast to the conception of the human race by scientists and politicians in the national socialist arena, the analysis of human alleles shows that the genetic variation between individuals within each race can be greater than the variation between races (Pääbo, 2001). Thus, there is a large probability of two humans within the same population differing more in their genetic traits than two humans who belong to different races. This is a fundamental correction of the view of race as it was commonly understood in the past. The realization that human beings can be more different within a race than between the races was most notably described by the Italian population geneticist Cavalli-Sforza. One of his important books, published in Italian, was released in Germany under the title "Verschieden und doch gleich" ("Different yet still the same") (Cavalli-Sforza and Cavalli-Sforza, 1994).

The ethical importance of the consequences of this state of affairs in population genetics led to the passing of a declaration on the issue of races among humans in 1995 in Stadtschlaining, Austria, which preceded the UNESCO conference "Against racism, violence and discrimination." This workshop was covered in the German magazine "Biologie in unserer Zeit" (Kattmann, 1996) in an article entitled "Vielfalt der Menschen, aber keine Rassen" ("Human Diversity, but No Races").

Indeed, races are products of human intuition reinforced by cognitive presets in our mind. It is wrong to consider individuals who differ little as belonging to the same race; those who are more different as belonging to different races; and those who are even more different as different species. This incorrect idea has been abolished by the modern population genetics findings that intra-racial differences can be greater than inter-racial ones.

Races are not populations of individuals who are different in a majority of their traits, but rather, they are populations of individuals who possess a few characteristic traits that are the diagnostic traits of their race (Chapter 5). It is easy to find differences in outward appearance (e.g., skin and hair color, morphology of the body and the face) between humans from different parts of the Earth.

However, the necessity of adaptation to particular environmental conditions in different geographical areas has only led to changes in a small number of genes in the genome, specifically those that have been subjected to the adaptive pressure of adjusting to local environmental conditions. In humans, the diagnostic traits between races are mainly related to skin and hair color as adaptations to the varying intensity of solar radiation and allelic differences in specific metabolic enzymes as adaptations to varying diets (Kingsley, 2009). The majority of genes in the genome are not affected by these adaptations, and accordingly, the numbers of these changes vary among single organisms within a race to the same magnitude as between members of different races. Hence, a native Central European can differ from his equally Central European neighbor in more genetic traits than from a sub-Saharan African. Races are not generally different from each other, but they carry specific "license plates," by means of which they can be identified. This realization disproves a common prejudice about the nature of racial difference.

The problem is that the differences between races concern very distinct adaptations to local environments, and these traits are altogether low in number. Who decides which traits are awarded the rank of being a distinguishing property of a race? What makes skin and hair color so special? There are hundreds of other traits that could in principle be used to divide a species into diagnosable groups. Why do we consider native Africans to be a separate race, but not the Bavarians in contrast to the Westphalians in Germany? Both of these populations are diagnosable groups that are distinguished by certain specific traits.

These considerations make it clear that races are typologically defined, and the traits chosen in considering a population as a race are a matter of highly subjective human decision-making processes. Though races are useful tools for pragmatic classifications, they are nevertheless artificial classes formed into groups by the human mind. Races are pretended to be biological units, but in fact they are only scientific or social or political constructs. No natural law defines races as cohesive natural groups.

The example of racial discrimination shows most clearly the fatal consequences human intuitions can face if they are not verified by scientific and logical foundations.

2.12 Species Pluralism: How Many Species Concepts Exist?

A remarkable number of species concepts are currently in use (Mayden, 1997). This multiplicity demonstrates, above any other point, that all of these concepts must be fictitious. These concepts must be human constructs and cannot represent entities that exist in nature. It is highly improbable that more than twenty different concepts all correctly describe the species that exist in nature.

It is questionable whether the current pluralism of many different species concepts is desirable or sustainable. The various types of entities that are termed "species" differ in their basic regularities are therefore not comparable (Ghiselin, 2002). The impression arises that there is not likely to be a definition of "species" that is able to simultaneously combine the characteristics of gene flow, descent cohesion and trait equivalence in an equally justified manner (Atran, 1999). For this reason, it is amazing that there are consistent (i.e., monistic) field guides available for the identification of particular animal and plant groups.

Mayden has itemized the 22 species concepts that are currently in use (Mayden, 1997) and established that many of these concepts are mutually incompatible. No chemist would accept the existence of different concepts of a chemical element that depended on the author, especially if these concepts were irreconcilable.

If there are different species concepts, how does one know when to use each concept? Not every organism with a different set of traits is a different species, and not all species have markedly different traits. This means that a species cannot be defined by traits. For example, female Mallards (*Anas platyrynchos*) are phenotypically very different from male Mallards, so why are male and female Mallards not different species? A female Mallard is closer in appearance to a female Pintail (*Anas acuta*) than to a male Mallard. Therefore, appearance alone cannot be a criterion for differentiating species.

An alternative definition of a species that is not based on trait similarity would be as follows: a species consists of organisms that mate with each other. Organisms that do not mate with each other belong to different species. Male and female Mallards mate, but mallards do not regularly mate with Pintails. This example shows that a markedly different definition of species, based on mating, has to be applied to account for the fact that male and female Mallards belong to the same species. The species concept of trait similarity and the species concept of reproductive compatibility are two different species concepts that cannot be licentiously combined. How does one know when to apply each species concept?

Mayden presents the 22 species concepts in great detail and with commentary, so the species concepts will not be repeated here. Instead, it is noted that most species concepts can be subsumed under three main concepts (Mahner and Bunge, 1997):

1) **The phenetic species concept**: This concept is understood as classification based on common traits. Therefore, the species is an entity of maximum covariation between existing and missing traits. The phenetic species concept is also termed the numerical species concept because it quantifies the differences in traits between the taxa and then performs the taxonomical classifications and species delimitations, a process that is now assisted by computers (Sneath and Sokal, 1973).

   The objections for a phenetic species concept are twofold:

   (1) First, the phenetic concept is a purely formal, operational concept that treats biological species as a group of organisms with similar traits. The phenetic species concept is not concerned with the biological processes that have led to

the formation of the species (de Queiroz, 1999 5323). The concept does not conceive of groups of organisms as natural entities (Hull, 1997).

(2) Secondly, there is an infinite number of characteristics that can be labeled as different traits between two organisms (Chapter 4). In the phenetic concept, there are no criteria regarding which traits are biologically relevant and which are not (Dupré, 1999). A major objection to the concept is that traits are so heterogeneous that any consideration of quantifying degrees of resemblance is impossible (Ghiselin, 1997). How similar must two organisms be so that they are "similar enough" to belong to the same species (Hull, 1999)? No pheneticist can answer this question. Accordingly, many authors think that phenetics is not a science (Mahner and Bunge, 1997). Mayr considered phenetics to be, at best, a starting point for further taxonomic analyses (Mayr and Ashlock, 1991).

The species concept of phenetics is not very different from the species concept that underlies the barcoding approach (Chapter 4). Both concepts ignore biological processes and only look for differences. However, barcoding is based on DNA sequence differences, whereas phenetics mainly considers phenotypic differences. This is an important difference because it protects barcoding taxonomy from being confounded by convergent (parallel-evolved) trait similarities. However, beyond this one point, barcoding taxonomy is a type of phenetics.

The phenetic species concept is the only species concept that can be equally applied to all phyla of living organisms and thus offers the broadest applicability (Dupré, 1999; Hull, 1999). It is a very simple concept that can easily be applied in practice because it avoids conflicts with controversial opinions regarding descent or gene flow relationships. Perhaps a modest confinement to pure phenetics would have protected taxonomy from becoming an eternally disputed "never-ending story" (see below).

2) **The cladistic species concept**: The cladistic species concept is that of descent and therefore of evolutionary kinship (Hennig, 1966). Species are groups of organisms that have the same common ancestor. Every organism that has a common ancestor (i.e., organisms are related to each other) belongs to one species. In strong contrast to the phenetic species concept, organisms are not grouped according to trait similarities but are defined according to their genealogical relationships to each other (Peters, 1998). The cladistic species concept does not recognize a change in traits along an evolutionary line. Alterations in traits (anagenesis, see Chapter 7) do not give rise to new species (Figure 2.3). If a group of organisms does not bifurcate, it cannot become a new species because determining a change in traits is liable to subjective standards of perception and evaluation. In contrast, bifurcation is considered an objective fact (cladogenesis, see Chapter 7).

The unresolved problem of the cladistic species concept is that all organisms on Earth are of common descent. Where does one species end and another begin? All *Felidae* (the cat family) possess a common ancestor, just as all Lions (*Panthera leo*) ranging from South Africa to India have a common ancestor. Why is the Lion a

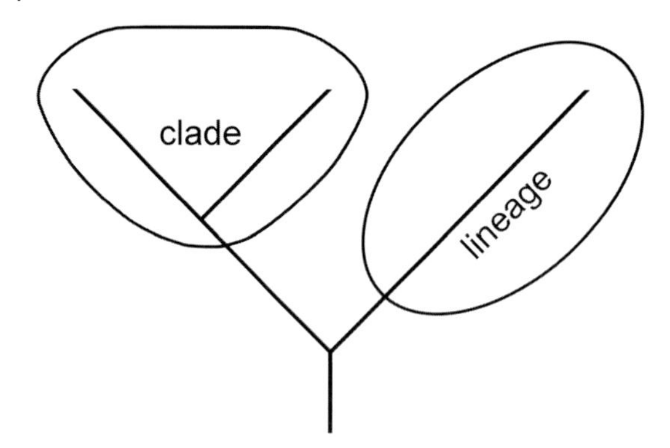

Figure 2.6 The phylogenetic tree consists of two elements: the lineage and the bifurcating clade.

species but not *Felidae*, which is instead considered a family? Both groups of organisms have common ancestors so this cannot be what makes them different. A common ancestor cannot define a species because families also have a common ancestor. Why are all organisms not just a single species?

Where are the limits drawn? A taxonomy based on common ancestry (monophyly) faces the problem of defining beginnings and ends (Mallet, 1995). When does a branch begin and when does it end? Until what point is a species a species, and when is the ancestor distant enough for the group to be a genus? In addition, evolution proceeds with different speeds. There are evolutionarily younger and older species; one cannot simply declare that an evolutionarily older species is a genus and that organisms with even older common ancestors are families, orders, and so on. Cladistics alone cannot define a species; it needs to borrow some of the classification principles of other species concepts to define a species.

Cladistics attempts to resolve this problem with phylogenetic bifurcations. The branches of the phylogenetic tree fork repeatedly, and there are lineages and clades (Figure 2.6). A species concept is defined according to this basic pattern of a phylogenetic tree, termed the "cladistic species" (de Queiroz, 1998). Therefore, a species is simply a lineage. As long as the branch continues as a lineage throughout the propagation of generations, the species continues to exist. However, as soon as the branch splits, the life of the species ends, and two new species (daughter species) begin which live on until new branching events stop their existence. With every bifurcation, two new species begin, and with every additional bifurcation of the two daughter branches, each of the two species ends and a new species begins. This appears to be a simple definition.

However, the entire notion of a cladistic species concept changes with the definition of a bifurcation. We certainly cannot define the species as the part of a branch between two bifurcations in the phylogenetic tree without first defining a bifurcation. The definition of a bifurcation (split) is very difficult because each birth of two siblings is a bifurcation from the parents into two daughter branches. Two brothers or two sisters differ with regard to many traits; the creation of a filial generation (F1) from a parental generation (P) is, of course, a cladistic bifurcation because the filial generation consists of individuals that differ in traits. What, then, is a taxonomically accepted bifurcation? When are offspring that are simply sons and daughters born, and when are species born? The definition of a cladistic species is not easy, if at all possible.

Every step of reproduction in which the parents produce more than one descendant is a cladistic bifurcation; only clonal reproduction is not a bifurcation into trait-differing offspring. Any other form of reproduction produces dissimilar offspring. Cladistics has attempted to resolve this problem by assigning a special rank to certain trait differences. Still, what justifies the award of a rank? This is exactly the weak point of cladistics; certain traits are counted as apomorphies, whereas others are not (Chapter 7), and the selection is subjective. Apomorphies are traits of a special rank that are awarded by which taxon differences are defined, whereas other traits do not define taxon differences.

Apomorphies are newly derived traits in the course of evolution. Autapomorphies are traits that distinguish two sister branches from each other. However, what exactly are autapomorphies? An autapomorphic trait is defined as one that defines the split into two different daughter taxa. But what exactly is the split into two different daughter taxa? The answer will be that this is a bifurcation of the evolutionary tree, defined by the appearance of new autapomorphies. Therefore, the matter is founded on circular reasoning.

From these results, it is clear that the cladistic species concept is a trait-oriented species concept, fraught with all of the subjective assessments and rating standards on which a trait-oriented taxonomical classification is also founded (Chapter 4). The acknowledgment and assessment of apomorphies is a purely subjective matter that relies on cognitive faculty, the assessment of quality, and the assessment of the extent to which organisms are different. Why is a feather, by which birds can be recognized and distinguished from reptiles, more important than other traits by which birds only marginally distinguish themselves from reptiles? Overall, the entire apomorphy-based bifurcation is a relapse into phenetic taxonomy.

However, there is the further possibility of defining a cladistic bifurcation from a stem group into two daughter species. It is also possible to define the split as the loss of the lateral gene flow connection among the organisms within a community (Chapter 6). If the gene flow within a group of organisms is disrupted, then a stem group splits into two separate daughter groups. However, even this definition does not support the species concept of a cladistic species; the species as a gene-flow community is a markedly different species concept than the species as a genealogical continuum of subsequent generations (Hey, 2001). There is no method of defining a cladistic species solely from the species criteria of its own concept. The cladistic species can only be defined by mixing this species concept with other species concepts.

We are faced with the problem that the aim to define a species by the criteria of cladistics always runs into a system of mixed classification criteria, including some that depend on subjective evaluation. No approach to conducting a species classification considering several principles simultaneously is capable of even-handedly considering trait equivalence, cohesion by reproductive gene flow and

a few organisms that do not possess one or several of these traits. In this case, the conclusion that must be drawn is that the identification traits that have been used thus far are inadequate, if not outright wrong, and the traits used for identification must be altered or extended. However, it would be wrong to conclude that these newly discovered organisms with other traits must be a new species.

In many cases, this incorrect conclusion does not lead to incorrect results because only those species that have been sufficiently well-researched are usually available to the field biologist for identification; for this reason, it is improbable or even impossible that the established identification traits are inadequate or wrong. Many species, however, are inadequately understood, meaning that a newly discovered group with different traits may well belong to the same species. Until now, the scientific world has been inadequately informed about the range of variation in some organisms, which can spread over the entire breadth of the species.

The diagnosis of a species is not the same as understanding what a species is as it exists in nature (Sterelny and Griffiths, 1999). The evidence for the presence of a new species cannot be concluded from the evidence for trait differences, just as the presence of one species cannot be concluded from the presence of certain trait identities, as there are many species with (almost) identical traits ("cryptic species") and that there are many trait differences across organisms that belong to the same species (polytypy). Thus, you can never say that organisms that are (almost) identical with regard to traits naturally belong to one species, whereas those with different traits belong to a different species. If you already know that two groups of organisms belong to different species, it is useful to be able to distinguish these species with the help of identification traits. If, however, two groups of organisms differ in certain traits, they are not necessarily different species.

The equivalence of traits is not what a species is, but it makes it possible to identify a species. The latter is an "operational definition," not a definition that tells us something about the essence of a species (Mahner and Bunge, 1997). A true definition (in contrast to an operational definition) should be able to describe what an object is, not simply what can be performed with the object or how to identify it.

An example of an operational definition may be: "What is a knife? A knife is an object that can be used for cutting." This sentence does not tell us what a knife is but only the things that can be performed with a knife. Another example is: "What is daytime? Daytime is a time on Earth when it is bright." This example does not tell us what daytime really is, only which trait allows us to recognize daytime. A solar eclipse, a volcanic eruption or a heavy rainstorm can make the day as dark as the night. Equally, lightning or floodlight can illuminate the night to be as bright as day. It cannot be correct to say that daytime is defined by brightness, just as nighttime cannot be defined by darkness. Brightness and darkness are traits for daytime and nighttime. However, they do not tell us what daytime or nighttime is.

One could say, "Well, it is usually true that if it is bright out, then it is also daytime, and if it is bright at night on one occasion due to some natural phenomenon or illumination by an artificial floodlight, then this is certainly realized and the spontaneous impression that it was day would be immediately corrected." However, defining daytime by the trait of brightness is wrong. The definition is wrong because daytime is not what "brightness" specifies.

Correctly, daytime (as opposed to nighttime) is defined by the position of Earth relative to the sun, not by its brightness. Daytime is the period of time at which the surface of the Earth is turned towards the sun. Nighttime is the period of time at which the surface of the Earth is turned away from the sun. In contrast, brightness is only a symptom of daytime, and darkness is a symptom of nighttime. These definitions are operational definitions. Bright light can be present even at nighttime. However, it can never be daytime in one location if the sun is positioned on the other side of the Earth. Summer and winter on Earth are defined by the inclination of the Earth's axis, not by their traits of being warm or cold.

Exactly the same type of reasoning applies to the biological species. The goal of identifying a given individual as a member of a given species should never be confused with the goal of knowing that a given group of organisms is a species. Nothing expresses the difference between diagnostics and ontology better than a statement by the classical author of evolutionary theory, American paleontologist George Gaylord Simpson (Simpson, 1961): "The definition of monozygotic twins ... provides a homologous causal sequence. ... Two monozygotic twins are not twins because they are similar but, quite the contrary, are similar because they are twins."

This quote expresses in a direct manner that on one level, there concern is about identifying twins. On another level, the quote addresses a markedly different issue, which is that twins are twins because they originate from a single egg. Twins have identical traits as a result of being twins, but this is not the ontological definition of twins because even non-twins can have identical traits. The fact that there are many field guides on the identification of species on the market, all of which focus on species diagnostics, may mislead us into falsely believing that diagnostic differences are ontological differences.

2.14
The Dualism of the Species Concept: the Epistemic vs. the Operative Goal

How can we finally solve the "species problem," meaning the worldwide dissension about what a species is? Different authors mean different things by the word "species" (Mallet, 1995). For more than a hundred years, this has led to what is called the "species problem". Disagreements about what a species is have led to deep dissent, so much so that it is referred to as a "never-ending story."

Taxonomists place varying weights on trait differences between organisms, descent or sexual cohesion between organisms. Why are there approximately 9000 bird species on Earth (del Hoyo, Elliott, and Sargatal, 1992)? It would be easy and consistent to defend other estimates, resulting in as many as 27 000 species (Cracraft, 1997). The number depends entirely on the weight given to particular delimiting criteria. There are many diagnostically easily recognizable subgroups, distinguishable races, genetically far-distanced groups or allopatrically isolated populations that would double or triple today's internationally agreed upon number of approximately 9000 bird species.

The situation is the same with regard to descent relations or kinship. Monophyletic, apomorphy-based bifurcations can be traced by modern molecular technology into the minutest branches, and in doing so, these traces can be justified in a theoretically consistent manner to represent speciation according to the cladistic species concept. There are always some autapomorphies if two sister individuals are born by the parents (Chapter 7). In the exploration of the cladistic bifurcations in the genealogical tree, branching into repeatedly new lateral branches can lead to an inscrutable chaos of clades. One branch follows the other and the branches are continually inter-nested (Mishler, 1999). This may increase the species numbers currently in use by a 100-fold (Cracraft, 1997; Mallet, 1995). Where does one begin to assign organisms to species, and where does one stop?

A similar situation arises with regard to species as gene-flow communities (Chapter 6). Tiny, isolated, reproductively separated populations can disaggregate into hundreds or thousands of groups of organisms that no longer have gene flow among themselves. These delimited gene-flow communities can be small if the organisms are not very mobile and if numerous barriers splinter the totality of organisms into the smallest separated gene-flow communities. Considering each isolated gene-flow community as a separate species also could increase the species number currently in use by a 100-fold.

What is the consequence of this dilemma? From this dilemma it follows that there are, and probably always must be, two different taxonomies (Schmitt, 2004). Biodiversity can be sorted into manageable units with which humans pursue goals and achieve results that satisfy practical needs, but nothing more. The second taxonomy is based on the theory of evolution. Descent and kinship play a role, as do reproductive communities. This is a taxonomy that does not follow operational convenience, but it satisfies the human need for consistency in lines of thinking.

The species problem as a "never-ending story" does not seem to stem from the disagreement about the more than twenty species concepts currently on the market (Mayden, 1997), ranging from the phenetic species concept and the concept of the reproductive community to the ecological species concept. This abundance of species concepts is usually called "species pluralism" (Hull, 1999). Instead, the species problem seems to hearken back to the fundamental difference between the theoretical concept of and the practical instructions for how to distinguish different species (Ereshefsky, 1999).

It may be unavoidable to separate both taxonomies from the start. Both taxonomies pursue different goals. One taxonomy conducts a classification that is feasible and serves for communication between scientists and laypeople. The other taxonomy is concerned with the biological unit that plays a role in evolution and is subject to natural selection for this reason. The question is to what extent nature has an "interest" in organizing organisms into cohesive groups that are maintained as isolated groups without merging into others. This kind of a natural species is a unit in biodiversity that would be present without human-imposed principles of order.

Practical taxonomy sees the species as a unit that satisfies a practical need. Practical taxonomy is concerned with manufacturing artificial classes into which the diverse individual organisms can be arranged according to their trait similarities or genetic distances. It is about a classification principle that makes the vast material of biodiversity utilizable and ensures communication between practicing taxonomists. However, this taxonomy is based on sorting principles that humans have shaped according to their own needs. The criteria of delimitation (e.g., of traits) obviously do exist in nature, but the results of the grouping, which are species, do not necessarily exist in nature. These species are made, not discovered, although the traits by which the organisms are combined into species were obviously discovered at some point.

Species, as they truly exist in nature, only require cohesive connections by which the organisms are joined. True species do not need to be sorted because they are already grouped in nature before human discovery. It is very important to understand this difference; many taxonomists do not see themselves as the inventors of species but as their discoverers.

Philosophizing about the strange world of taxonomy is not relevant to the daily routine of science. "We work with all of this, without really having to think about it," is what some taxonomists say. If you ask them more probingly, taxonomists commonly fall back to the pragmatism, "At the end of the day, we can work well with this."

Repeatedly, what the biophilosopher David Hull clearly expressed in his article "The ideal species concept – and why we can't get it" (Hull, 1997) seems to apply. Either the species concept is theoretically conclusive, consistent and based on the natural laws of evolution and is thereby unsuited for practical taxonomical applications, or the species concept orients itself according to operational pragmatics and is neither strictly consistent nor founded on a reproducible law. The biological species as a real unit is not suited for the everyday life of a taxonomist or mutual worldwide communication because of a lack of practical manageability. Conversely, the biological species as a workable unit for the division of biodiversity in practice is unacceptable for the theorist because of a lack of consistency and prevailing contrariness. The more consistency is sought, the less the classification of species is practical (Hey, 2001), and the more practicality is sought after, the less the classification of species represents consistent thinking. All of these factors lead to a dualism in the species concept.

The best example of this dilemma is the hundreds or thousands of attempts to replace the typological taxonomy with a consistent monophyletic taxonomy (Chapter 7). A consistent monophyletic taxonomy cannot work in practice and did not become widely accepted until present day. Attempts have been made to completely replace the Metazoan system with a new system consisting only of monophyla (Ax, 1995). However, the past two decades have shown that all attempts to replace classical taxonomy with a strictly monophyletic taxonomy have failed because they were not widely accepted. Practicing taxonomy does not tolerate theoretical consistency.

Another example of the rejected efforts to introducing theoretical consistency into practicing taxonomy is the futile attempt to replace Linnaean nomenclature with the "phylocode" (Chapter 7). Because the genus names in Linnaean nomenclature convey the wrong impression that the genera among different animal groups are comparable in rank, there was an effort made to replace the Linnaean names with the "phylocode" (Cantino *et al.*, 1999). Although this would have been an important step towards consistency and unambiguousness in taxonomy, this reform in nomenclature has not yet had the chance to become widely accepted. The only reason for this failure is that the reform is impractical.

It is telling that recent reforms in taxonomy have all sought compromises. The barcoders must always pronounce that they do not ignore other kinds of taxonomies and that these taxonomies are valuable supplementations. The "phylocoders" have realized that their new taxonomic names will not be accepted, and they seek a compromise by retaining the old names and explaining them using the new names. This situation seems to show nothing else than that taxonomy is a science that is inherently incompatible with consistent ways of thinking.

Perhaps it was a mistake from the beginning to include phylogenetic branching and the concept of reproductive coherence in the practice of taxonomy. If taxonomy restricted itself to what it originally was, which was a method of sorting according to sensibly selected principles, then most of the discrepancies that taxonomy has encountered since Darwin would disappear; however, taxonomists would also have to consistently commit themselves to this method. The goal of taxonomy would then need to be pragmatism, and pragmatism in most cases means a classification according to diagnosable differences in traits.

To end taxonomy's fate as a "never-ending story" and an eternal point of contention, it seems that separating the goals of pragmatism-oriented taxonomy and causal science is unavoidable. Remarkably, this idea has already been suggested by Ernst Mayr, a persistent advocate of a nomologically oriented taxonomy, who recommended the complete severing of the species as a nomological unit from the routine business of a taxonomist, who is forced to classify organisms into groups (cited in Mallet, 1995). Subsequent authors have again and again arrived at the same conclusion (Heywood, 1998; Atran, 1999; Schmitt, 2004). Even Ghiselin (2002) agrees with the proposal of giving different names to classes of species for the purpose of operational handling and the species as it exists in reality.

The natural and the artificial species concepts both have a right to exist and will both be used side-by-side. Many disagreements could be avoided if this dualism of the species concept was admitted and if the pragmatist would admit that "his" species units are artificial products. The unmasking of this difference explains the substantial amount of heated emotions concealed behind the species problem. The species problem is about a mutual lack of comprehension, mutually different esteems by other disciplines of natural sciences and varying successes in the fight for appreciation as a high-ranking science. There are reasons why taxonomy faces a fight in competition with other branches of the natural sciences. The range of taxonomy extends from "folk biology" ("scientia amabilis") that is close to science in our everyday lives to a natural science that is based on reproducible and falsifiable experimental results, in agreement with the philosopher Karl Popper.

3
Is the Biological Species a Class or is it an Individual?

3.1
Preliminary Note: Can a Species have Essential Traits?

The question regarding whether the species unit is a matter of human sorting principles or a unit that really exists in nature is repeatedly asked in this book. This question relates to whether the species is viewed as a class or as an individual.

To approach the problem more closely, one can ask, "What is it that makes an organism belong to a species?" An organism must certainly be characterized by something, the results of which designate it as a member of a species. Is an organism a member of a species because it has a particular descent, because it has sexual connections and therefore lateral gene flow to other organisms (Chapter 6) or because it has particular traits? It is surely not simply a Tiger's colored stripes that make it a Tiger because even a Tiger without stripes is a Tiger just the same. It is crucial to ask the question, "why is a non-striped Tiger a Tiger?"

Accordingly, is there a particular feature that makes a Tiger a Tiger? This would be the Tiger's essence. If such an essence was to exist, then the following would be true: (1) If one particular organism did not have the Tiger's essence, then it would not be a Tiger. (2) Every organism that possessed the Tiger's essence would have to be a Tiger. The essence of a species would therefore be something that is necessary and sufficient for an organism to belong to a species (Sober, 1994); does such an essence exist in the case of species?

Beginning with Aristotle on to Linnaeus, Putnam (1975) and Kripke (1980), there was no doubt that species essences existed, and all of these philosophers and scientists believed that these essences were intrinsic traits of the members of a species. The conviction that there are essential intrinsic species traits resulted in biological species being considered universal, similar to an element in chemistry. Furthermore, just as an element possesses an essence, which is the number of protons it has, the biological species was also believed to possess an essence, namely, its species-specific traits.

Gold has 79 protons. If a particular element does not have this number of protons, then it cannot be gold, and every element that has 79 protons must necessarily be gold. The number of protons an element possesses is something that is necessary and

Do Species Exist? Principles of Taxonomic Classification, First Edition. Werner Kunz.

sufficient for the element to belong to the class of this element. This type of logic applies to chemical elements.

However, is there a trait in an animal species that every organism must necessarily possess to belong to the species? Would every organism on the planet need to belong to this animal species if it had the trait? The answer is no (Wilkins, 2010). Biological organisms have no traits that are necessary and sufficient for taxonomic membership. Every trait that characterizes the species membership of an organism can be absent from each individual organism, and if such a trait that characterizes the species membership of an organism is absent, this organism nevertheless does not lose its species membership. A non-striped Tiger is still a Tiger.

This is very surprising; the fact that the absence of a species-specific trait does not preclude membership of the organism to a species clearly shows that traits cannot be the essence of species membership. Why is the number of protons in a chemical element a trait that does not tolerate exceptions, in contrast to the traits of living beings that do tolerate exceptions? The essence of species membership must be something other than traits (see Section below: "The relational properties of the members of a species are the essence of the species").

Darwin's theory of evolution explains the fundamental differences between the essences of chemical elements and living beings. Darwin put an end to the idea that animals or plants have essential traits. The intrinsic traits of biological organisms cannot be essential for these organisms' affiliations with a species because all of the biological organisms' intrinsic traits are subject to mutative change thus evolution. Neither morphological, physiological, ethological, chromosomal nor genomic traits are essential for the members of a species and neither are certain DNA sequences. This is because all intrinsic traits can change permanently, including the DNA sequences in the genome. In this respect, there is no difference between a DNA sequence and a phenotypic trait. Both are diagnostic of the species, but neither is essential. If a species-specific DNA sequence mutates and thus changes, the affected organism does not lose its species membership. If one were to assume that a species is only that which has a very particular set of traits, then a species could, in principle, only exist for a single moment because some traits may already have minimally changed by the next moment. Therefore, a species cannot be defined by its intrinsic traits.

Species are subject to reproduction and continuing mutative changes. In chemistry, the elements do not multiply but remain constant. No gold atom divides into two daughter gold atoms and then dies; this means that the objects of chemistry are present as rigid classes. The chemical elements have always possessed this quality, and they are likely to continue to express this quality in the far distant future. The elements have properties on Earth that are identical to their properties on distant planets. The elements are universals in the philosophical sense (see below), at least in the frame of our currently existing, spatiotemporally restricted universe.

The objects of biology are an entirely different matter; they change constantly, and it is impossible for even one of our animal or plant species to exist simultaneously on a distant planet. If there are animal or plant species on distant planets, they must be completely different species or they must have migrated there from Earth.

3.2
Class Formation and Relational Group Formation

Organisms can be grouped in two fundamentally different ways. First, one can sort according to trait resemblance, which is a class formation and a grouping based on equality of types. Equal- or similar-looking objects are combined into a group, resulting in the problem of whether groups like these are accurate because humans can, of course, choose the traits they want and form groups that are based on any trait resemblance (Devitt, 1991). Given a number of objects that have a mixture of several traits, one can form a considerable number of different groups from that number of objects, depending on which trait you select to base group formation on. Because it is certainly possible to combine objects according to the purely subjective criteria of human preferences, it is at least feasible that groups are formed that do not exist outside of the human need for sorting.

Groups of objects with equivalent traits are called classes (here, the concept of a class should not be confused with the usage of the word "class" as a category in taxonomic classification alongside the categories of genus, family, order, etc.). Another alternative consists of combining objects into groups if the single objects are bound to each other; this means not simply bonds that are generated by conceptualizations in the human brain but objectively existing links that would exist even if there were no humans. Group formations such as these are not based on similarities in intrinsic traits but on relational cohesions. These formations are based on the organisms' relationships to each other; for example, a red item belongs to the class of red objects because it has the property of being red. A neighbor, however, is a neighbor because he possesses a certain relationship to the person of reference, not because he possesses a certain intrinsic trait that would make him to be a neighbor. Class formation and relational group formation are two entirely different grouping methods that are based on different conceptual laws thus they may not be mixed (Ghiselin, 2002).

These insights are of enormous importance for taxonomy because taxonomy is group formation, and it is critical to ask from the beginning whether individual organisms are sorted and grouped according to traits or whether they are combined into a group because they have certain ties to other members of the group.

The traits by which class formations can be made are intrinsic traits, meaning those that the organisms carry entirely within themselves. Every member of the group carries all components required to make it a member of its group in and of itself; a second red object is not needed to make the statement that a particular red object belongs to the red group. Upon seeing a single object that is red, it is immediately clear that it must belong to the class of red objects. To belong to a certain class, it is sufficient to have the intrinsic trait that unites all group members. The members must have this trait, but nothing else is required.

An entirely different matter is the relational group, in which the members of the group must be cohesively or causally connected to other group members (Figure 2.5). An additional object of reference is always required in relational grouping. Upon seeing a single man, you immediately know that he belongs to the male group, not to the female group. This is an example of a class formation. In contrast, if you see a single man, you cannot tell whether he is a neighbor. At a minimum, you need a second individual to make the statement that the first individual belongs to particular groups. A brother needs at least one sibling to be considered a brother. An outside observer cannot tell whether an individual is a brother or a father by his intrinsic traits; information about his relationship to other people is needed (Bock, 1989). In contrast to class formation, in relational group formation no group member carries the criteria for group membership within themselves.

The difference between a class formation and a relational group formation can be explained with a simple example: if many children are playing in an area, and you want to divide them into groups, there are exactly two options, as explained above (Figure 2.5). First, you can combine the children into groups by traits. For example, you can (mentally) combine all of the boys and consider them separate from the group of girls. This would be a classification by a trait, leading to two different classes (classes *1* and *2* in Figure 2.5). As a completely different method, there is a second grouping alternative. You can unite all children into a group that holds hands with each other and are separated from children in a second group; this is a type of relational grouping that leads to two groups, which are both entirely different from those in the first example given (relational groups *1* and *2* in Figure 2.5). The second method of grouping is based on entirely different principles of affiliation; it is a grouping according to mutual cohesion. For one particular child to belong to the group, it is not enough that the child has any one particular trait within itself; the child must have a (real) relationship with a second child.

The difference between class formations, as opposed to relational group formations, also becomes clear in the following example: the term "man" leads to an entirely different group affiliation than the term "brother." "Maleness" is an intrinsic trait that assigns every man to the male group. This group affiliation is given to every individual man by his own account and does not require an organism of reference. You do not need a second man or a woman to belong to the male group. However, "brotherhood" is a concept that necessarily requires a person of reference, namely, a second sibling, to meet the requirements for the term "brother" and to belong to the brother group. One human being on his own can certainly be a man, but he can never be a brother on his own. A person of reference is also necessary to be a daughter; otherwise the concept of daughter does not make sense. In this case, the person of reference is a mother or a father. Without a mother or a father the concept of "daughter" cannot be applied meaningfully.

These considerations are of an immediate relevance to taxonomy. Each taxonomist should be aware of the foundations for group formation because taxonomists make or discover groups. Can an organism belong to a species if it exists by itself? Does it inherently carry all of the criteria for a species affiliation? Does the organism need a second organism of reference to belong to a species? The answer to these questions depends entirely on the method of group formation that is applied by the taxonomist, that is, either class formation or relational group formation. This example alone makes clear how important it is to first understand the foundations of group formation before taxonomical classifications are conducted.

3.3
Is the Biological Species a Universal/Class or an Individual?

The primary question addressed here is whether biological species are universals, meaning a group that can occur more than once in the world (Armstrong, 1978). A class is always a universal. A group of large objects, for example, can occur more than once in the world. Applied to organisms, the question is whether any animal or plant species can be a class; if so, can the species occur in more than one location and time in the world? This concept must be explained more precisely.

An animal species is considered to occur more than once in the world if it can occur at different locations in the world at the same time and if these occurrences are independent. A species would only be a class if its individuals can occur multiple times simultaneously. These multiple occurrences must be ontologically independent from each other. This concept is not intended to refer to individuals of the species that have spread from the origin of the species. Instead, it means that the individual occurrences have nothing to do with each other. The biological species would be considered a universal or a class only if this condition is fulfilled. House sparrows (*Passer domesticus*) would be a class if the house sparrows of North America were independent from the house sparrows of the Old World, but this is obviously not the case.

It is important to understand the meaning of the phrase "can occur" at different locations, which is not the same as "does occur." The logical principles of class formation allow multiple and independent occurrences of classes at different locations and at different times across the globe, but a prerequisite must be that the particular objects that are being classified are indeed present several times. Class formation does not dictate that the members of a class must exist repeatedly but only that when particular objects exist repeatedly they can be grouped into a class because these repeated occurrences are independent. The concept of class formation only comprises the logical rules of one of several ways of how to group diverse objects.

Let this be explained by the following example: an animal species, for example the Tiger, can only be a universal or a class if the Tiger could occur at location A and also at location B. A class is a universal, and universals occur in the world without being spatiotemporally restricted. Only the objects of a class, the single occurrences of the universal, are spatiotemporally restricted. These single occurrences are called particulars or instances. These particulars are individual occurrences, and they are ontologically independent from each other. The class is not an individual occurrence. A particular piece of gold on Earth and a piece of gold on Mars are instances in which their occurrences are independent from each other. It would be wrong to assume that a piece of gold on Mars would not exist if a piece of gold on Earth does not exist. Therefore, gold is a universal.

Now, back to the Tiger. Presently, there are no indications that Tigers exist on any distant planet but this possibility cannot be excluded. To understand the concept of class formation, we simply assume that Tigers exist on a distant planet. If this was the case, would it then be possible to group Tigers into a class? No, because we know from the laws of evolution that all biological organisms do not originate *de novo* and independently from each other but instead by descent. The Darwinian theory of descent implies that the discovery of a Tiger that shares all of the properties of a terrestrial Tiger on a distant planet must be interpreted as a migration of the Tiger from the planet Earth to a distant planet, or in turn from the distant planet to the Earth. If the Tiger has migrated to the planet from Earth, then it is not independent from the terrestrial Tiger but part of the group of Tigers.

Therefore, these assumed distant Tigers cannot be independent elements of a class but rather parts of one single whole. A descent community cannot occur simultaneously and independently at different locations, so the species as a group of organisms that are bound to each other cannot be a universal. A descent community cannot be a class.

If, however, these assumed Tigers on the distant planet are independent occurrences (i.e., the result of a *de novo* origin but nevertheless identical to terrestrial Tigers in all of their traits), then these so-called Tigers would be the result of convergent evolution. However, under the conditions of parallel evolution these apparent Tigers on the distant planet would not be considered Tigers.

This fundamental difference between chemical elements and living beings is based on the fact that chemical elements are not subject to evolution. Elements arise in the stars; they are not born by parents, and they are not subject to mutation and selection. Tigers cannot arise by astrophysical processes. Tigers only propagate an entity that already exists. This realization dates back to Louis Pasteur who stated that life arises from preexisting life, not from non-living material. This awareness is summarized in the phrase *Omne vivum ex vivo*, Latin for "*all life [is] from life*," also known as the "law of biogenesis." As a consequence of the law of biogenesis, living beings cannot be universals or classes.

The opposite of a universal or class is the individual. The perception that biological species are individuals has been founded mainly by Michael Ghiselin (Ghiselin, 2002). Here the term "individual" is used in the philosophical sense, meaning an entity that is unique and exists only once in space and time. The concept of an individual should not be confused with the use of the word "individual" in everyday life, in which the term "individual" refers to a single organism, not to the totality of the group.

An individual can only occur once in the world. In contrast, a class is always a universal. A class is a set of objects with some coincident traits. A group of objects is always either a class or an individual; it cannot be both simultaneously, nor can a class and an individual be mixed. The difference between a set of objects as a class and a set of objects as an individual is a fundamental one. As an individual, the species "Tiger" is perceived in an entirely different way ontologically than it would be perceived as a class. As an individual, the Tiger's physical properties do not affect why all Tigers belong to a group. The physical properties cannot define group affiliation of the members of an individual. A single Tiger only belongs to the individual "species Tiger" because it has a relational connection to other organisms, namely, a descent relation (Chapter 7) or a gene-flow relation (Chapter 6), not because it has particular properties. Of course, the individual "species Tiger" has a number of physical properties, but these do not define why particular organisms belong to the individual "species Tiger" nor is the individual "species Tiger" defined by physical properties.

Ultimately, the biological species cannot be a class. Of course, trait-equivalent organisms can be grouped into a class, but (if applied to biological organisms) this procedure would contradict the laws of evolution. Each attempt to unite organisms into classes leads to artificial constructs that are fictive units that do not exist in reality.

3.4
The Difference Between a Group of Objects as a Class and a Group of Objects as an Individual is a Fundamental One

A group of objects is always either a class or an individual (Ghiselin, 1997; De Sousa, 2005). A group can never be both a class and an individual at the same time, nor can class and individual be mixed. These differences can be summarized as follows:

1) **The individual exists in space and time**: An individual always occupies a region of space and exists at a certain time. Because an individual exists at a certain time, it always has a beginning and an end. At a certain time in the past it did not exist, and at some point in the future it will no longer exist. An individual is a historically fleeting occurrence. When the individual has become extinct or has been exterminated, it cannot reoriginate. The Eiffel Tower has a founding date, and one day it will no longer be there. A newly constructed, similar-looking tower at the same or different location is no longer the Eiffel Tower. The Polar Bear (*Ursus maritimus*) originated approximately 200 000 years ago (Breiter, 2008), and due to the current climatic change, pessimists predict its early demise.

   In contrast, a class is not restricted to a region of space or a particular time of existence. A class cannot begin to exist and it cannot become extinct. Because a class is a group of several organisms based on common properties, a class can exist everywhere, every time, provided that the objects of the classification are present. As long as the objects are there, it is inconceivable that a class would exist at only one point in time. If a class exists at a particular location, then it can also simultaneously exist at a different location in the universe. Likewise, a class might exist once at a particular time but then again later at a second or third time. The multiple occurrence in space and time must not be realized in our world, but no contradiction arises if this concept is assumed. The only requirement is that the objects with similar traits are again present at other locations or at other times. In every imaginable world there will be a class of red objects, provided that red objects exist there, and in the distant future there will always be a class of red objects. Any claim that the class of red objects had become extinct would be meaningless.

   Let us apply this concept to the Polar Bear. Of course it is possible to construct a class into which the existing Polar Bears fit well. This class would be a group that combines a number of properties such as white fur, other morphological traits, and species-specific behavior. However, this is an artificial group. Whenever there are animals on another planet or at a different geological time that could fit into this class, one can apply the class concept to these organisms without any problems or contradictions. However, the class as a group in itself is a fictive object that does not exist at a certain location or in a certain time. The class is a universal. However, as an individual organism that is formed by common descent and by gene flow cohesion, the Polar Bear exists in reality. The Polar Bear exists at a certain location and in a certain time; it arose at a certain point in time, and it will disappear at a certain location or at a certain time.

2) **The individual consists of parts, the class of elements or instances**: A class consists of elements that have equivalent traits and are independent from each other. These instances or elements of a class are neither cohesively connected to each other, nor do they causally or ontologically depend on each other with regard to their existence. If you remove one red ball from the class of red balls, the other balls are not affected; beyond identity or similarities in traits, the red balls are unrelated.

   However, an individual as a whole consists of parts, not elements. In contrast to the elements of a class, the parts of an individual are connected to each other. These connections may be (1) cohesive bonds, (2) causal dependences or (3) ontological cohesions. Cohesively connected signifies that one is physically linked to the other. Causally connected signifies that one is the cause of the other. Ontologically connected signifies that one would not exist or not make sense if the other wasn't there.

   (1) The Eiffel Tower is an individual (not a class of metal elements) because its iron parts are physically connected to each other.
   (2) A particular thunderclap is part of a particular, individual thunderstorm (and not an element of a class of meteorological events) because thunder is the causal consequence of the electrical discharges that occurs during a thunderstorm. Lightning and thunder are mutually dependent.
   (3) A football team is an individual (and not a class) and its players are the team's parts because the goalkeeper cannot be a goalkeeper without the other players. The goalkeeper would lose his meaning without the team. The removal of one player affects the entire team. Even if the football team goes separate ways for a time, the goalkeeper does not lose his existence as a part of the team as long as there are the other players on the team and the team does not decide to permanently end their cohesion. The existence of every part of an individual depends on the existence of the other parts of the individual.

   Let us apply this to the Polar Bear. If you remove one Polar Bear from the group of Polar Bears, this does not have a large impact on the other Polar Bears. However, in principle, it has an influence. The Polar Bear that is removed cannot reproduce or further influence the other polar in any other way. In principle, all other Polar Bears are affected, at least to a minor extent.

   The two vertically arranged groups of boys (left) and girls (right) in Figure 2.5 are examples of a class formation. At all times, one can assign a group of children to classes of girls or boys. The two horizontally arranged groups, however, are individuals because the children mutually touching are cohesive groups and their cohesion is a historically fleeting occurrence with a beginning and an end.

   The difference between the parts of an individual and the elements of a class can also be seen from the fact that the parts of an individual can be just about indefinitely cut up into small and even smaller pieces while still being parts of the individual. A single nail of the Eiffel Tower is certainly a part of the Eiffel Tower. In contrast, the elements of a class can no longer be divided; they constitute an ultimate size (Mahner and Bunge, 1997). It would be absurd to call a child's pullover an element of the class of children.

3) There are no natural laws for individuals, only for classes of individuals (Ghiselin, 1997) because laws are universal statements (e.g., "All Xs are Ys."). Natural laws are statements that are always true at all locations and at all times. A law can always be applied to more than just one case. While a law can affect a single individual, it also always affects other individuals because it has universal validity.
   In the real world, natural laws are, in principle, independent from space and time. As long as we do not leave our actual world and consider parallel worlds, the law of falling bodies, for example, is always true, everywhere. It is not possible to say that the law of falling bodies was only true at one particular location or at one particular time but not at other locations or at other times. While there are certainly laws regarding the celestial bodies as a class, there is no natural law on our Milky Way as an individual group of stars, just as there is no natural law regarding the group of all minerals, of which Mount Vesuvius is composed, as opposed to Mount Etna. If Mount Vesuvius consisted of only a single class of minerals, then there would obviously be natural laws for these minerals; however, then these are the laws of the class and not unique rules about Mount Vesuvius's minerals. Water is a class, and laws about water always refer to water as a universal that is independent from space and time, not to a group of all water molecules in a particular glass that is filled with water.
4) Classes always have defining properties, whereas individuals cannot be defined (Ghiselin, 1997). You can define why a particular object belongs to a particular class; for example, the property of redness defines why a red object belongs to the class of red things. An individual, in contrast, cannot have any defining properties. You cannot define why a part of the Eiffel Tower belongs to the Eiffel Tower. Any one of the Eiffel Tower's iron lattices has no properties that define why this lattice is part of the Eiffel Tower. The concrete planet Earth, our Milky Way and the man Darwin cannot be defined. Of course, Darwin has certain properties, but these properties do not define why Darwin is Darwin. An individual that is equivalent with regard to all of Darwin's properties would still not be Darwin. Individuals can only be assigned a designation: Eiffel Tower, Earth, the Milky Way, Darwin. However, you can certainly define what a planet, a galaxy or an evolutionary biologist is because these are class concepts. The property of being definable distinguishes classes from individuals.

In a superordinate sense, individuals can be grouped into classes. The grouping of all imaginable football teams into the class of football teams is possible. This kind of group would then be a class, which can also be seen from the fact that this group formation is not spatiotemporally restricted. All imaginable football teams in the world cannot become extinct or be exterminated. If all of the teams were physically annihilated, they could regenerate. The class of football teams certainly has defining properties; the number of individual elements (always eleven), the footwear and other clothing, the objective of the game (to score goals) and so on. A team that consists of twenty players and pursues a different aim than making goals would be an element that does not belong to the class of football teams, but the group of eleven players on an individual football team is an individual, not a class. "Manchester United" cannot be defined.

For example, if one needed to define the Great Tits (*Parus major*) by their physical properties, they would just describe an individual by its properties. The Great Tits have black head plates, yellow bellies, perform a particular song, and so on, but this is not a definition of Great Tits; instead, something that has an individual existence is described, not defined.

The only plausible way to consider a terrestrial Great Tit as an element of the class of Great Tits would be if there were other groups of Great Tits on other planets that shared Great Tit properties. However, the theory of descent implies that the discovery of a bird on a distant planet that shares all of the properties of a Great Tit must be interpreted as a migration of Great Tits from the Earth to another planet or in turn from the distant planet to the Earth. However, then these distant Great Tits would not be elements of a class but rather parts of one single individual. Otherwise, these apparent Great Tits on the distant planet would not be Great Tits but would instead be the result of an independent convergent evolution. Considering these animals to be Great Tits would be an incorrect view that ignores the biological rule of descent. Hence, the fact that biological species are individuals, not classes, directly follows from the theory of evolution.

Repeatedly, there have cases in which an animal species that has already been registered to be extinct is then later rediscovered, usually in remote places rarely visited by humans. Of course, we all assume that the newly discovered remaining members of the species had a relational connection to the extinct population. Therefore, they are a part of a whole that was once been widespread. No one would nurture the hope of rediscovering an extinct animal or plant species as a *de novo* existence. Classes, however, cannot become extinct because they are universals.

3.5
Artificial Classes and Natural Kinds

A class is a group of objects with coincident traits. This leads to the problem, depending on the selection of traits, of every imaginable aggregation of objects into classes is possible. It only depends on a human's decision regarding which traits he would like to use for group formation. Imagine one hundred objects that each has one hundred traits, resulting in a large number of possibilities for combining all or some of these objects into classes. All of these possible class formations cannot be realistic classes that exist in nature. Some of these classes would have to be pure nonsense. Which selection of traits leads to a meaningful or even correct class formation?

Some possibilities for selection are noticeably subjective because they reflect nothing but personal taste, such as the class of ugly looking worms (Ghiselin, 2002). Other group formations that are made according to trait resemblance are suspected of being purely human-made. These are the class formations that are based on traits that are objectively measurable in nature; for example, formations of items that have the same color are grouped into a common class: the class of red objects. Such classes may be assumed to exist in reality, though the question of reality is disputed here (Riggs, 1996). There are divergent opinions about whether a class of red items really exists or whether these classes are also figments of the human mind that only serve communication with fellow human beings (Goodman, 1956; Armstrong, 1978). The author of this book supports the view that a class of red objects is not a naturally preset entity but a mental construct although, of course, the red objects are natural.

The correlation of class formation with natural causes or laws is crucial for the question of whether classes are natural or human fiction. There are obviously artificial classes that serve the human need for order and mutual communication, and these are obviously not discovered but made. Such artificial classes do not exist outside of our minds. Only the members of the class exist, but the classes themselves do not exist as natural groups (Devitt, 1991). For example, an artificial class would be the group of all objects whose names begin with a "D." Charles Darwin, the extinct Dodo and all ducks form a common class according to this criterion. Without a doubt, the class of all German citizens that were born on May 1st 1990 is an artificially made class.

The class of bearded men is also an artificial class because no natural law is broken or disregarded if a member of the class shaves and becomes be excluded from the class (Ghiselin, 1997). Only the human principle of order forbids the continued treatment of the shaved individual as a member of the class of bearded men. However, the principle of order is not a natural law; the group of bearded men is an artificial construct that does not in reality exist in nature.

However, there appear to be classes that are not the result of a human sorting effort because their class cohesion is based on natural causes or laws. These classes are called "natural kinds" (Riggs, 1996). Although natural kinds are classes and are hence based on trait equality, like the artificial classes, and not on relational connections, like the relational groups (see above), they are not made by humans.

To make this difference more clear: It is important to distinguish between nominalistic and realistic classes. Although both are groups of objects whose group affiliations are based on trait similarity or equality, the former are artificial constructs whereas the latter are natural. Nominalistic classes are linguistically or mentally constructed aggregates of objects that are connected by nothing other than the fact that they fall under a certain predicate or concept (Kripke, 1980; Putnam, 1975; Riggs, 1996). In contrast, natural classes ("natural kinds") exist independently from our language and our brain activity and would even exist as groups without humans to sort them. Realistically understood natural classes are groups that are based on natural laws. The trait similarities that unite the members of such groups in natural kinds permit nomological generalizations.

Natural kinds are based on trait essentialism (Wilkins, 2010); essentialism refers to the principle of "necessary and sufficient" (see above). A trait is the essence of a class if every member of the class has the trait, without exception; otherwise, this member would not belong to the class. Conversely, if there are any objects in the world that possess this exact trait, then this object would necessarily belong to this class.

For example, all gold atoms form a class of this kind. Gold has 79 protons. Every atom in the world with the atomic number 79 must necessarily be gold. Gold is sufficiently defined by its atomic number of 79. If gold were to lose one proton and thus obtain the atomic number 78, it would no longer be gold but platinum. Conversely, there can be no gold in the world that does not have the atomic number 79. If a particular element does not have 79 protons, then it cannot be gold, and every element that has 79 protons must necessarily be gold. An element's proton number is something that is necessary and sufficient (i.e., essential) for the element to belong to the class of this element. All gold atoms in the world thus form a natural class type. This is a "natural kind." From a common trait, the atomic number 79, general nomological statements can be derived that make gold what it is.

When in 1869 Dmitri Mendeleev and Lothar Meyer determined the nomological connection between the number of protons (atomic number) and the chemical properties of the individual elements and accordingly constructed the periodic table of elements, they discovered a natural type of affiliation among the chemical elements. In other words, they discovered that the chemical elements belong to classes that are natural kinds. Before this discovery, the elements were grouped in artificial classes by combining them according to properties such as gaseous, metallic or reactive (see below).

In summary, it can be said that there are two different groups with different ontological structures, the individual (1) and the class (2):

1) A group of cohesively, causally or ontologically connected objects is an individual. It is a group of relationally linked objects and is real. It is a historically transient occurrence, and because it is unique, it is an individual.
2) A group of objects with equivalent traits and whose connection consists only because they share one or several traits is the class. The class is always a universal; its existence is not spatiotemporally restricted. It can exist at several locations in the world independently, and it can exist several times in the world; these existences are independent of each other. Classes fall into two different concepts: they can either be nominalistic constructs, that is, human-made artificial classes, or they can exist as realities, that is, natural kinds.

3.6
The Biological Species Cannot be a Natural Kind

Every attempt to classify biodiversity into units by their trait similarities leads inevitably to units that are human-made constructs. However, it is evident that when classifications are undertaken according to traits, they mean something else. Traits are used, but the intention is to classify by relational, descent or gene flow connections (Figure 2.4).

There is a fundamental difference between what a species is and how a species is recognized (Chapter 2). Usually, the traits used for species identification are not theoretically justified. It only matters that they are as reliable as possible for distinguishing one species from another. This, however, assumes that the existence of species is a given.

The aim to identify a species fulfills an entirely different ontological purpose than the intention to specify what a species is. If a species already exists, then it can be identified as exactly this species through its traits, which do not require any theoretical justification. However, if it is important to justify why a group of organisms is a distinct species in comparison to another group, such a question cannot be decided by traits because there is no justification for which traits are admissible. The decision about whether something is a species or not can only be made through the species as a relational group. Only those that are a relational group can be a species. A group of organisms with matching traits is always an artificial construct unless there are other reasons that it is a species, and then the traits are only useful to identify and delimit them as such.

The history of taxonomy reveals that different traits were used for taxonomic classification in different times, thus creating taxa that are no longer acknowledged as such. In the seventeenth century, bird classification was mainly done by beak shape and foot structure. For example, the Loons (*Gaviidae*), the Grebes (*Podicipedidae*), the Coots (*Fulica* species) and all species of *Anatidae* (Ducks, Geese and Swans) except for the Screamers (*Anhimidae*), were united in the Swimming Birds group (Chen, 2002). Swimming Birds are a class of organisms that share a number of traits and thus constitute a biologically meaningful grouping. All have webbing between their toes, can move relatively elegantly along the water's surface without danger of drowning, and all Swimming Birds have the special ability to effectively grease their plumage, thereby making it impenetrable to water.

Currently, there is not a class of Swimming Birds in contemporary biological systematics, neither as a family or order of birds. Every taxonomist that takes this realization as trivial should ask himself why a taxon of Swimming Birds is no longer accepted in contemporary biology. This answer is at the root of the matter because this example of class formation leads to an artificial unnatural group. Why is this group unnatural? All traits that led to this group's formation are certainly very natural and lead to a very biologically meaningful group. In this case, it certainly cannot be said that absurd taxonomical grouping traits were chosen.

The premise of combining all swimming birds into a common taxon follows very natural biological reasoning for forming a group. Which arguments led to the conclusion that a taxon of Swimming Birds should be discarded as a group that is no longer biologically meaningful? The answer to this is unambiguous and elucidating; there are no arguments of any kind that rest on trait similarity. Instead, the taxon of Swimming Birds has been discarded because most of the similarities between *Anatidae*, Loons, Grebes and Coots are cases of convergences. In other words, the Swimming Birds taxon has been discarded because the organisms were united in the taxon that were not related to each other. As a consequence, a trait-based grouping does not have to lead to natural taxa with any certainty because this is unreliable. Only relational connections lead to natural taxa, that is, the descent or gene-flow community.

These considerations show that it is possible to group biological organisms according to trait similarities. However, the groupings achieved are not units that exist in nature. Natural units are only achieved if the taxa are based on the relational connections of organisms. A species cannot be defined by its traits alone. There is something more fundamental behind it, which is often only intuitively perceived.

Traits are not the essences of biological organisms. There is no species in which only a single intrinsic classification trait would satisfy the conditions of an essential trait, that is, a trait that could classify all members of a species. Which criterion makes a robin belong to a group of robins? Could an essential trait be a subtler criterion, such as the DNA sequence in the genome? Kripke (1980) and Putnam (1975) defended the organisms' genome as the species essence and took the view that certain DNA sequences were necessary, sufficient and thus essential for an organism's species affiliation.

However, they were incorrect. In biological species, there are no species-specific DNA sequences in the sense that their alteration would instantaneously cause a species' affiliation to be lost. Genomic DNA sequence traits are also not necessary and sufficient for an animal or plant to be a member of a certain species. Accordingly, every attempt at grouping biological organisms into classes based on intrinsic traits has remained inherently incomplete, which is why biological species cannot be natural kinds.

3.7 The Biological Species as a Homeostatic Property Cluster

An alternative approach to understanding the biological species as a natural kind is offered by Boyd (1999), who represents a different view of the natural kind that he calls "homeostatic property cluster." A homeostatic property cluster, in contrast to the traditional natural kind, does not require that the class be defined by essential traits. Instead, it suffices to understand the class as a group of elements that share several stable similarities. These similar traits that the class members (elements) have in common do not need to be present in all members of the class; they are not essential. However, they must be stable enough for their presence in class members to be more than coincidental. It must be possible to predict, with better than chance probability, that the members of the class have these properties.

For example, Common Swallowtails (*Papilio machaon*) have many traits in common. They have black-yellow patterned wings, tail-like elongated extensions on their hindwings, red and blue spots on their hindwings (Color Plate 1), and their pupae are held in an upright position by a silk girdle. The theory of understanding the members of the species Swallowtail as a homeostatic property cluster does not require that every single Swallowtail has black-yellow patterned wings, tail-like elongated extensions on the hindwings, red and blue spots on the hindwings, and pupa that are held by a silk girdle. There occur some exceptional rare mutant individuals that do not have some of these properties. All of these traits are not essential for species membership, but one can predict that a particular Swallowtail actually has black-yellow patterned wings, tail-like elongated extensions on the hindwings, red and blue spots on the hindwings, and pupa held together by a silk girdle with better than chance probability. All of the traits together form a stable property cluster.

There are three biological reasons for the stability of these properties. First, the members of a biparental biological species exchange their genes with each other as a gene-flow community, which keeps the members of a species fairly homogeneous over a reasonable time. Second, all members of a species have a common ontogenetic developmental program because they all originate from common phylogenetic ancestors. Third, all members of a species are susceptible to common selection conditions because they live in the same environment and thus remain fairly constant and homogeneous. If these adaptive traits were to deviate beyond a certain limit in an individual Swallowtail, this particular individual would have reduced fitness and would thus have a lower chance of surviving and reproducing. Natural selection would keep anomalies within a certain limit. However, this is only true to a limited extent, and it is often not true in the case of geographically distant races.

However, homeostatic property clusters are clearly different from essential traits. Traits can only be valuated to be essential traits, if all members of the group share these traits without exception. The occurrence of each trait must be necessary and sufficient for the members to belong to the group. Furthermore, homeostatic property clusters define not only biological species but also other biological categories, such as genera, families or orders, and are likewise defined by homeostatic property clusters:

1) The higher taxonomic categories also share common properties that can be predicted with better than chance probability. For example, the felids (cat-like animals) share the common property of being able to retract their toe claws, unlike the members of other carnivorous families. The mammals share the common property of having body hair, as opposed to birds and reptiles, but felids are a family, and mammals are a class. The definition of a species as a "homeostatic property cluster" therefore is not a species concept at all, but a generally admitted concept for categorizing biological taxa.
2) The phenomenon of convergent evolution is another example that the criteria for a homeostatic property cluster defines several other biological categories and is not at all restricted to the species category. Convergent evolution means that non-related organisms are controlled by the same selection pressures and, thus, can look very similar in a number of traits without belonging to a single species (see several of the Color Plates). Groups of convergently evolved organisms form homeostatic property clusters, with all of the characteristics for the better-than-chance predictability that Boyd demands for the species as a "kind."

The theory of the species as a homeostatic property cluster attempts to save the biological species as a "kind," but this approach runs afoul of Darwin's theory of evolution. Evolution causes permanent changes in the traits of organisms, and evolution also causes permanent changes in homeostatic mechanisms. The members of a species therefore do not only change their individual traits but also change their homeostatic mechanisms within a species. Across time and geographical distances, species members are often susceptible to varying homeostatic mechanisms.

3.8
Polythetic Class Formation or Grouping According to Family Resemblance

Biological species do not possess a single trait that is present in all members of the species. It is simply not true that there are genetic properties that all members of a given species have in common and that all members of a different species lack (Okasha, 2002). A human has 23 chromosome pairs, while our closest primate relative, the chimpanzee, has 24. However, genetically defective humans with supernumerary or subnumerary chromosomes, such as those with chromosome 21 trisomy and affected by Down syndrome, are without a doubt still human. The chromosome number of a species is not an essential trait for the species.

There is always some trait that individual organisms of a species lack, and in spite of this, these organisms still belong to the species. This is one of the arguments for the conclusion that the biological species as a class of organisms cannot be defined by essential traits and, therefore, cannot be a natural kind (see above). Advocates of a trait-based species concept for class formation do not have arguments for why an organism belongs to a class if it does not have a species-specific trait.

However, it can be argued that only a few traits have to be used simultaneously to define class affiliation. It can be argued that, although the members of the class do not share a single common trait, they share several traits. Under this condition, class affiliation is not assigned by any essential single trait, but instead by the possession of several different traits. Different classes are then distinguished by a multitude of attributes that do not all have to apply simultaneously for a particular class element. Different members of the class in each case have several, though not all, traits in common. Thus, a species is characterized by clusters of covarying traits, not by the possession of any common essential trait.

The biological species as a cluster of covarying traits can be illustrated by the following scheme. The five organisms *a* through *e* all belong to the same species. Each organism is characterized by four traits. However, not all of the five traits *A* to *E* are shared by each of the five organisms:

Organism *a* has properties *A B C D*.
Organism *b* has properties *E B C D*.
Organism *c* has properties *A E C D*.
Organism *d* has properties *A B E D*.
Organism *e* has properties *A B C E*.

A class like this is a polythetic class as opposed to a monothetic class, in which a single factor determines whether a particular organism belongs to this and no other class (van Regenmortel, 1997). Wittgenstein (1953) called this phenomenon "family resemblance." An example for a polythetic class from another field would be a disease that is characterized by a number of symptoms. Not all symptoms have to appear simultaneously; in each single patient, a single symptom can be absent, but the patient still has this specific disease.

Thus, if a biological species can be defined at all by traits, then it can only occur in the sense of statistically covarying traits (Hull, 1997). One particular Northern Pintail (*Anas acuta*) is not a Northern Pintail because it necessarily has a pointed tail but because it also has other traits and behavior patterns that delimit it from, for example, the Mallard (*Anas platyrhynchos*). The common possession of several similar traits justifies the classification of many individual organisms as the species "Northern Pintail" and distinguishes the Northern Pintail from the Mallard.

However, although a polythetic class is clearly based on a similarity of traits, no single trait is essential for class affiliation. Therefore, the polythetic class cannot be a natural kind. No single factor determines whether a particular organism belongs to this and no other class. No single trait is necessary and sufficient for class affiliation. Class cohesion is not given due to natural laws. On its own, family resemblance does not support a nomological generalization and is thus something entirely different than the natural kind of a chemical element due to the number of protons of its atoms.

3.9
The Linnaean System is Based on Fundamental Assumptions that are Irreconcilable with a Contemporary Worldview of Science

From Aristotle to Linnaeus to certain modern authors, the idea of the biological species has always been that of a class. Organismic diversity was grouped according to trait similarities. The best-known representative of the viewpoint that the biological species is a class is Linnaeus, who grouped the diversity of living beings into a system by combining organisms with similar traits into taxa. Linnaeus had the conviction that these taxa would really exist in the world, and he saw his task as that of discovering them. Thus, Linnaeus was in no way a pure utilitarian or pragmatist whose primary concern was to divide organismal diversity into manageable groups. In contrast to modern pheneticists (Chapter 4), he was convinced he was discovering reality. Linnaeus believed in the existence of groups. He perceived the biological species as a real unit and not as a template created by humans into which the existing diversity was allocated. With his viewpoint, Linnaeus followed Aristotle's traditional conception. It is important to note that there is a fundamental difference between the Linnaean species concept and today's phenetic species concept. Linnaeus believed in the existence of species as natural kinds, while pheneticists consider their (likewise) trait-oriented classification to be merely a pragmatic concept, not something truly existing.

Linnaeus believed in traits as being essential because he believed that these traits would be sufficient and necessary for a species' membership in a class. The Linnaean way of thinking assumed a divine act of creation with which the species were created in the beginning and have remained more or less unaltered. Linnaeus imagined the species truly existed in the external world, independently from our mind (Ereshefsky, 1999). Because of the traits' constancy, he did not encounter any difficulties in assigning organisms to rigid classes. Linnaeus's taxonomic worldview had not yet been "tarnished" by the findings of Darwin, who discovered that all traits were subject to evolution. The theory of evolution implies that all traits are fleeting with the logical consequence that a rigid allocation of organisms into classes is not possible. There cannot be classes if all elements undergo continuous change.

Linnaeus's species concept was that of a natural kind, in which members are characterized by essential traits. To Linnaeus, species were types in the platonic sense and had an autonomous existence. Behind the sum of individual organisms, Linnaeus imagined the species to formally exist in the external world and not only in our mind (Ereshefsky, 1999). Linnaeus considered species and genera as units existing in nature, and he based his binary nomenclature on this worldview. But only species and genera were real to him. Higher taxonomic categories, such as families and orders, were perceived as artificial constructs created by humans only for pragmatic reasons, even by Linnaeus (Ereshefsky, 1999).

The Linnaean nomenclature is still in use all over the world. This is astonishing because the Linnaean classification system was only possible under ideological assumptions that are no longer scientifically accepted. Species essences cannot be reconciled with the theory of evolution. It would be expected that the Linnaean species concept would have been overthrown by Darwin's findings. The idea of the species as a type cannot be reconciled with variation due to evolution. According to Linnaeus's species concept, logic would force us to conclude that an individual organism would immediately lose its species affiliation if it loses an essential trait.

Linnaeus assigned the organisms to groups according to the resemblance of their traits. Individuals with significant similarities were combined to "species," but Linnaeus could not arbitrarily use all traits for his classification. He had to look for criteria, and some traits were "suitable," while others were not. Of course, according to Linnaeus, the body size of an organism was not a property that could be used for taxonomic classification, because it was immediately clear to him that body size can vary. If body size was of taxonomical importance, then slightly larger and slightly smaller organisms would not be allowed to belong to the same species. Linnaeus also knew that the blossom coloration of many plants was not a reliable trait either because it can depend on the plant's age and on the pH of the soil.

Linnaeus's criteria for the "suitability" or "unsuitability" of traits were based on their stability. Linnaeus intuitively chose only a few from a large number of traits for taxonomical classification, namely those that were not altered much by environmental influences, such as weather, climate and location. (Here, "alteration" does not mean evolution, but environmental change.) Linnaeus classified the stable traits as taxonomically important and the variable traits as taxonomically unimportant, and he saw the taxonomist's task as that of distinguishing between essential and non-essential traits to use only essential traits for species classification.

He classified flowering plants according to the "sexual system," that is, by male and female sexual organ traits, which are closely linked to sexual compatibility. By doing so, Linnaeus intuitively came very close to the natural species concept of the reproductive community (Chapter 6), even though he still conceived of the species as a class and not as a relational connection among organisms. However, with an uncanny instinct, he mostly included those trait groups that were correlated with the real species as a gene-flow community.

3.10
Comparison of the System of Organisms with the Periodic Table of Chemical Elements

The chemical elements differ from each other in their reactions. Some are reactive, while others are inert, and some can only enter into very particular bonds, while others are capable of a multitude of possible bonds. Both chemical elements and biological organisms possess a number of traits that allow classification into specific groups. At first glance, there appears to be no difference.

Accordingly, the classification of the chemical elements into classes originally did not differ in principle from the classification of plants and animals into classes. Those with similar or identical properties were combined into a class. This means that, due to insufficient scientific knowledge, the chemical elements were not recognized as natural kinds in the past but were instead (just as biological organisms today) classified according to human-made classification principles.

In 1869, the Russian chemist Dmitri Ivanovich Mendeleev and his German colleague Julius Lothar Meyer discovered, independently from each other, that there were physical laws hidden behind the elements' property classes. With this awareness, Mendeleev and Meyer created the periodic table of chemical elements, which is fundamentally different from a classification of animals or plants because the grouping of chemical elements is based on physical laws that could only be resolved by quantum mechanics at the beginning of the twentieth century.

The outermost electrons of an atom's shell are responsible for its chemical properties. The number of available valence electrons increases with atomic number; elements with only one outermost electron available for chemical bonds are assigned to group I, elements with two electrons available for chemical reactions belong to group II, and so forth up to group VIII. The arrangement of electrons in group VIII elements is especially stable; these are the noble gases, which are especially inert for this reason.

In the periodic table, the elements are ordered from left to right by increasing atomic number (number of protons) and thus by the number of available valence electrons. After eight elements in the horizontal row, an element that opens up a new shell with its valence electrons appears. It resembles an already classified element with regard to its chemical properties and is accordingly arranged below an already inserted element, which begins a new horizontal row in the periodic table. In the end, elements with similar reactive properties are located in vertical columns on the periodic table. These vertical columns are called "groups." The eight vertical columns express the law of octaves, which contains the elements' reactive properties.

With this presentation, the fundamental difference between the system of chemical elements and the system of biological organisms becomes apparent. A class of particular chemical elements in a group on the periodic table are far more than a class of objects with similar traits. There is a scientific justification for the trait identity that is based on a consistent natural law. The group of noble gases is not a polythetic class formation due to the coincidence of several properties, but rather a natural kind due to a physical law.

In contrast, the system of animals and plants is not based on natural laws that stringently dictate which selected organisms must be grouped into a natural class. The grouping of butterflies and moths into the order Lepidoptera due to their scaled wings and other traits does not include any law contained in the scaled wings and other traits. Every attempt to group biological organisms into classes has not lead to natural kinds. The only way to group biological organisms into natural entities is the combination of the different organisms by their relational cohesion. The natural law does not come from properties but from relational descent and gene flow cohesion, and for that reason, butterflies and moths form a natural group.

That the classes of the chemical elements are based on a physical law also follows from the fact that extraterrestrial planets and solar systems are composed of exactly the same elements as Earth because the same natural laws apply there. If "aliens" one day land on Earth, they will certainly be composed of the same atoms that we are, but even if their properties were identical to ours, they would never be humans. The only imaginable possibility for them to be humans would be if they were our relatives, that is, that they had traveled from Earth to the distant planet or from the distant planet to Earth some time ago.

3.11
The Relational Properties of the Members of a Species are the Essence of the Species

If biological organisms are grouped into classes according to trait resemblance, then these classes are not natural. However, because taxonomists apparently have no problem with assigning individual organisms to species, there is something more fundamental than trait resemblance that leads to the certainty of the existence of species (Davies, 2005; Okasha, 2002). Species appear as natural groups, but cohesion by mutual relational is what makes the groups natural. What is it that makes a Tiger a Tiger? It is not the striped fur but the connections with its ancestor and with its sexual partners.

Biologists and philosophers of biology typically regard species essentialism as incompatible with modern Darwinian theory. Samir Okasha, however, has shown that the standard antiessentialist considerations only show that species do not have intrinsic essential traits (Okasha, 2002). However, this does not mean that the biological species does not have any essential properties; relational properties are the essence of a species. If an individual Tiger did not share a common ancestor with all other Tigers, it could not belong to the species Tiger. All of its properties are rather unimportant; it only must belong to one and the same descent community.

The descent cohesion and gene flow relation between organisms makes an organism a member of its species. Any particular organism belongs to a species because its species affiliation depends ontologically on the existence of other species members. Every organism that is excluded from the gene-flow community and is therefore separated from the other organisms is not a member of this species. Vice versa, every organism that is part of the gene-flow community and is thus connected to the other organisms must belong to this species.

This realization compels one to accept that there is an essential species affiliation. However, it is not an essence based on intrinsic traits, but an essence based on relational connections. Why should the application of the philosophical concept of an essence be restricted to intrinsic traits only? Nothing compels one to restrict applying the concept of essence to traits only (Okasha, 2002). The integration into a gene-flow community is a necessary and sufficient condition for an organism's species affiliation.

Authors from Aristotle to Linnaeus to some more modern authors (Putnam, 1975 and Kripke, 1980) took the view that species essences had to be intrinsic traits of the species members, with the consequence of the biological species being treated as universal. Darwin's theory of evolution put an end to this view, after which species essentialism was called "untenable" (Dupre, 1999). Okasha (2002), on the other hand, has once more taken up species essentialism by breaking the connection between the concepts "essential" and "intrinsic trait." Instead, he presents the idea that species essentialism could be based on relational connections.

4
What are Traits in Taxonomy?

4.1
Preliminary Note

The laws of evolution determine how biological species originate, and these laws do not imply that the members of different species all must differ diagnostically to the human eye. The laws of evolution also do not imply that members of the same species have to be trait-identical (Ghiselin, 2002). Different species may be phenotypically very similar in traits or they may be very different (see several of the Color Plates). Similarly looking species are called cryptic species (Mayr and Ashlock, 1991).

The remarkable phenomenon of mimicry includes many examples for cryptic species (Wickler, 1968). Mimicry is the adoption by one animal species of the phenotypic traits of another animal species in order to be protected from predators. This phenomenon results in the evolution of parallel phenotypes between animal species that are not phylogenetically related with one another. If one species (the mimic species) is palatable to predators, whereas an equally appearing animal species (the model) is distasteful, the mimic will be mistaken by the predator for the unpalatable model. This scenario is called Batesian mimicry (Color Plate 2). If the individuals of two phylogenetically unrelated species are both unpalatable to predators, there is a selective advantage for the convergent evolution of similar phenotypes, because once the predator has tasted one individual, it will avoid both species in the future, since both have the same warning coloration and therefore will be protected. This scenario is called Müllerian mimicry (Color Plate 3).

Conversely to cryptic species, there are also polytypic species. Polytypic species are species whose members have distinctly different traits. Nevertheless, they are considered to belong to one and the same species because they are not reproductively separated from each other (Chapter 6). These few examples make obvious that it is impossible to define the biological species by its traits. The biological species does not possess a single trait that would be necessary and sufficient for species membership (Okasha, 2002).

Do Species Exist? Principles of Taxonomic Classification, First Edition. Werner Kunz.

The concept of "traits" assumes a central role in taxonomy. However, there are some serious problems in using traits for taxonomic classification:

1) First, traits are principally only acceptable for the diagnosis of species that are already known to be species. This often is not realized. In every case, the use of trait differences for the diagnosis of species implicitly requires that the species that shall be identified is already known to the scientific community as a species. Newly discovered individuals with new traits are not necessarily new species (Sterelny and Griffiths, 1999) (see Section "It is one thing to identify a species, but another to define what a species is" in Chapter 2). The importance of traits for the diagnosis of species is indisputable. However, diagnosis is not definition. The term "trait" has a very different meaning in taxonomy depending on whether it is used for diagnosis or for definition. A trait can serve as a *differentia* or as a *definitio* with respective fulfillment of different functions.
2) Second, most traits used in taxonomy are the phenotypic end products of the expression of various different genes. This is called polygeny. For this reason, most traits are confusingly complex building block systems, consisting of many single components that are coded by very different genes with different evolutionary origins and different homology relations to each other. Most of the traits used in taxonomy are collective terms for structures that are put together from several different single traits. These single traits each have their own evolutionary history. Most traits used in taxonomy cannot be homologous or non-homologous to each other for the simple reason that they are complex. Only the basic components of the phenotypic traits can be homologous to each other, not the traits themselves (see below).
3) Third, traits are always perceived and evaluated subjectively for taxonomical classifications, depending on the technology available, and the subjectively justified assessment of whether particular traits are important or not. Certain traits are ignored, others are respected, and nobody knows the rule behind this distinction. This means that traits are very dangerous if not useless for defining a species membership, although some modern taxonomists argue the converse (Wägele, 1995).

4.2
What Basic Rule Defines Traits as Being Taxonomically Relevant?

Higher organisms are diploid and thus always have two alleles of each gene in them. The two homologous alleles of the diploid set can be different. Allelic diversity in whole populations (multiple allelism) is even larger because each allele can display sequence differences tracing back to founding mutations that came into existence in the evolutionary past.

Because higher organisms have 20 000–30 000 genes in their genome, the differences between two arbitrary organisms can number in the tens of thousands. Admittedly, many of these differences do not lead to observable changes in phenotypic traits, but one can assume that there are many hundreds if not thousands of traits that distinguish every particular organism from others in a population. Every organism is unique in its complexity.

Every biologist is aware of this phenomenon. However, less commonly known is the obvious problem of how one can actually speak of species differences if the members of a species differ so drastically in regard to their traits. What is the actual difference in traits between species when all organisms within a species differ from each other? Only monozygotic twins are identical in traits.

The phenomenon of allelic diversity conjures the difficult question of what are intraspecific differences in contrast to interspecific differences. What is behind the fact that different species are distinguished from each other by their traits? Every user of an animal or plant identification guide uses the option of distinguishing one species from another species by certain traits. Is it the extent of differences that makes it possible to distinguish species, such as the principle that few differences equals belonging to the same species but many differences equals belonging to a different species? Or is it the quality of differences that makes it possible to distinguish species, such as the principle that certain differences equals belonging to the same species and other differences equals belonging to a different species?

Neither is the case. Numeric criterion can be eliminated immediately. It cannot be the extent of differences. Otherwise, males and females from a great number of animal species would have to belong to different species. Qualitative criteria can also be refused. The principle of "What 'only' differs in the number of bristles belongs to one and the same species, but what differs in the number of legs are different species," cannot be generalized either.

Every multicellular organism has many thousands of traits. Even today in the age of genome sequencing, in which the number of animal and plant species whose genomes are completely sequenced continually grows, there is no understanding of how many phenotypic traits a particular species possesses in contrast to another species. The logical conclusion is that the characterization of a species by its traits always has to be based on a restricted selection of traits and that all traits can never be considered. We presumably will be able to do this one day after the sequencing of all genes and all gene regulatory and synthesis pathways have been deciphered. However, this day still lies in the far distant future. Currently, every taxonomical classification is assigned according to traits that are based on the selection of a few traits. This selection is subjective.

The scientist, however, should have the desire to establish an underlying system of rules for general validity, which define the quality of traits suitable for taxon membership. Such a theoretical foundation for a trait-based taxonomy should define for all organisms a system of rules for which traits are relevant for taxonomic classification and which are not. However, this desire cannot be fulfilled.

The traits that distinguish one species from another are not quantifiable. It is not possible to determine a number of traits that are species specific. From this, it follows that a statement such as the following is senseless: "The organisms of two different populations match with regard to half of their traits." For example, Wilhelm Meise claimed 70 years ago that the Italian Sparrow (*Passer italiae*) was a hybrid species of the House Sparrow (*Passer domesticus*) and the Spanish Sparrow (*Passer hispaniolensis*) because it matches with both "parent species" in regard to half of all traits (Töpfer, 2007) (Chapter 6). This statement is scientifically worthless because the concept "half" results from a purely subjective selection of traits.

There are no fixed rules for trait-based taxon membership (Christoffersen, 1995). The selection is based on intuition, not on rules. This awareness opens the door for different taxonomists to use different traits to execute taxonomic classification. However, the choice of different traits for taxonomic classification might lead to different taxon entities. A flashback on the history of taxonomy shows that in different epochs different traits were used for taxonomic classification, creating taxa that are today no longer acknowledged as such. For example, bats were included in the taxon of birds from the antique naturalist Plinius to the Swiss biologist and bird painter Konrad Gessner in the sixteenth century because the property of having wings had been considered as an essential trait for being a member of the birds. Another example is the screamers of South America (genus *Anhima*). These were formerly considered to belong to the *Galliformes* (the order of pheasants, hens and quails) because they do not possess webbed feet. In fact, they belong to the *Anseriformes* (the order of ducks, geese and swans) because it has been discovered that the property of having webbed feet is not an essential trait for being a member of the *Anseriformes* (Chen, 2002).

The realization that, for many taxonomical purposes, traits are subjectively selected should not be misunderstood to indicate that the selection was arbitrary. They are not. Trait selection is subjective but not arbitrary because there is a distinction between taxonomically useful and less useful traits. Linnaeus already distinguished between the more variable and more constant traits and only chose the latter for his taxonomical classification (Chapter 3). Ernst Mayr also realized this. However, he justified his approach to the rating of traits by saying that with experienced taxonomists it was founded on "considerable knowledge and experience" (Mayr, 1982). Thus, he did not bring a rule into the world because he completely appealed to intuition. Mayr viewed this approach as a testament to taxonomy being fundamentally different than physics and other natural sciences. In doing so, he did not consider that with this analysis he labeled taxonomy and other disciplines of organismal biology as "soft science," potentially harming organismal biology (Chapter 2). This is not to be taken lightly because in the last half century a trench has developed between the biological disciplines labeled as "classical" and the biological disciplines labeled as "molecular," although these vocabularies do not resolve the underlying problem (Roush, 1997; Pigliucci, 2002).

In the 1930s, the well-known geneticist Alfred Sturtevant tried to develop a method with which somebody without any experience at all, even a non-biologist, would be able to divide species into "natural" groups. The essential element of such an approach would be the development of procedures with which the degree of resemblance could be quantified so that subjective taxonomy would be transformed into an objective, numerical taxonomy (Mayr, 1982). However, Sturtevant failed and did not realize his intention.

In recent times, numerical, computer-based phenetics has attempted this (Sneath and Sokal, 1973). Phenetics classifies organisms by only their trait similarities.

However, this taxonomy is a purely operational concept (Hull, 1997). There are strong objections to such a concept (Chapter 2).

Because phenetics treats biological species as simply a group of organisms with similar traits, phenetic classification results in taxonomic entities that do not reflect evolution (Dupre, 1999). The phenetic species concept does not consider any biological coherence whatsoever, neither common descent or sexual cohesion. It is a purely formal concept, which is far from recognizing units that have biological-evolutionary effects in nature.

Furthermore, phenetics aims to quantify the differences in traits between the taxa. Taxonomic classifications and species delimitations have been defined by the number of trait differences. However, this aim cannot be realized because there is an infinitely number that could be called differing traits between two organisms. Because phenotypic traits are so heterogeneous any thought of quantifying a degree of resemblance becomes impossible (Ghiselin, 1997).

The modern method of "barcoding" is also pledged to objectivity and tries to arrange species into "natural" groups. Barcoding is the ambition to classify all organisms into species solely by differences in certain DNA sequences, the so-called barcodes (Miller, 2007). This is an interesting parallel to the original claim by Sturtevant. But at what point are two groups of organisms distant enough from each other to be separate species? The barcoding method does not provide information about this (see below).

4.3 What is the Relevance of Differences in Genes Between Two Species?

It would be a very naïve idea that every trait corresponds to a gene. Genes are the information carriers and traits are their products. Indeed, classical genetics by Mendel and Morgan is based on this principle. Classical Mendelian genetic inheritance of gene passage to the offspring, gene linkage and recombination are based on the assumption that traits are the visible equivalents of the genes. The Mendelian rules are carried out with traits, but as a proxy for the genes. Particular traits, such as eye color or the curvature of a bristle in *Drosophila*, are visible, but their underlying genes are unfortunately invisible.

However, the concept of classical genetics should only be understood as a rough idea, which in most cases is incorrect with regard to the details. The one-gene-one-trait principle is only applicable in the rarest of cases. In most cases, the one-gene-one-trait principle is an idealization that is much more complex in reality. In reality, most single traits are controlled by many different coexisting genes (polygeny).

In most cases, whether a trait is present in an organism does not depend on the presence of the structural gene that encodes this trait. In many more cases, organisms of different species do not differ in their structural genes (Prud'homme *et al.*, 2006). A structural gene is a gene that governs the morphological development and the exterior appearance of an organism from the egg to the embryo to the adult. Body structure proteins usually consist of several hundred amino acids. In the case of humans and chimpanzees, almost thirty percent of these proteins are completely identical; in both species, they have the exact same amino acid sequence. The remaining seventy percent of the body structure proteins differ in only one or two amino acids (Enard and Pääbo, 2004). Why are the relevant genes and proteins the same but not the traits? Why does a human look different from a chimpanzee? This is a crucial question of taxonomy.

What is considered to be a single trait in taxonomy is a mosaic structure at the genetic level. In most cases, a trait is a multicomponent structure that is controlled and regulated by several genes. Most importantly, these genes have different evolutionary origins and are mostly not homologous to each other (see below). Whether a trait is present in an organism almost always depends on the structural gene's regulation. The gene can be switched on earlier or later in development. It can be switched on to a stronger or weaker degree. Large differences in morphology between two organisms, which may be of crucial taxonomic importance, can be caused by the same structural gene being switched on at different time or to a different extent.

The decision whether a structural gene is switched on depends on enhancers and transcription factors. These elements may have markedly different evolutionary origins and histories from the structural gene that encodes the respective protein. Enhancers are small DNA sequences that are assigned to particular structural genes. Enhancers are genetic switches operated by the protein products that are called transcription factors. Transcription factors are protein products encoded by still other genes that may be localized to different chromosomes, and they may have markedly different evolutionary origins and histories from the structural genes which they regulate.

Enhancers are switches that regulate the expression of their associated genes. In principle, every tissue has a different switch that is active or inactive, and this switch is regulated by transcription factors. A single transcription factor may regulate different structural genes in the same organism, and a single structural gene may shape the phenotype of different traits in the same organism. The same structural gene can be used again and again in the same organism, even in markedly different contexts, and so it may lead to the expression of many different phenotypes.

In animals, most of the genes for shaping the body structure regulate several different body traits (Abzhanov *et al.*, 2004). This is called a pleiotropic polyphenic effect. If such a pleiotropic gene suffers a mutation, many body parts are affected. However, this pleiotropy does not mean that the switching on or off of a gene in different tissues or organs has to happen everywhere in the body at the same time. Gene regulation is absolutely capable of switching on a single gene in different tissues at different developmental times and with different intensities.

Enhancer mutations can alter individual body parts and leave other body parts, whose development is controlled by the same gene, unaffected. In speciation, this path has obviously often been pursued when skeletal parts changed. By enhancer mutation, individual body parts can be altered selectively without any harm occurring somewhere else.

Against all earlier expectations, many genes and their proteins have an astonishing resemblance even among distant groups of animals. At first glance, one cannot distinguish whether a genome is from a mouse or human (Prud'homme *et al.*, 2006). For ninety-nine percent of human genes, there is a mouse counterpart. In the genome, fairly little has changed since approximately 100 million years ago. The coding sequences from different animals have mostly remained conserved. It is not the genes, which encode for structural proteins that cause animals to look so different. Here, the phenotypic appearance is deceiving. How an animal is built anatomically, which and how many body parts and extremities it develops, which size, form and color the body parts have, seems to lie predominantly in the modified expression of the structural genes and not in the genes themselves. Different traits between two different species do not result from differences in particular genes, but instead from the fact that the exact same genes are activated in a species-specific manner according to their local and temporal patterns of gene activity.

Coding sequences only make up approximately one and a half percent of the genome of animals. Everything else is DNA sequence with other functions or no function at all. Of this one and a half percent, less than ten percent is actually "body structure genes." The majority of genes are required for the everyday tasks that regulate basic cellular metabolism and have nothing to do with taxon differences.

These considerations make clear that it is misleading to attempt to find an analog for a trait pattern differing between two species (phenotype) in the gene pattern (genotype). The very hope that classifying taxa on the gene level rather than the phene level would be more exact is called into question. In some cases, a taxonomy based solely on the similarity of structural genes or on the origin or evolutionary pathway of these genes would result in a markedly different classification than the currently used classification of organisms.

Genetic differences between species are complex networks of only a minor number of genes. It is very difficult to find those genes that are really responsible for species differences. It is a naïve conception to consider species differences as nothing else than differences in a number of DNA base exchanges of arbitrarily chosen DNA sequences (see criticism on barcoding below).

4.4 In Sticklebacks (*Gasterosteus aculeatus*), a Single Gene Controls Many Phenotypes

The Three-spined Stickleback (*Gasterosteus aculeatus*) occurs in Europe, northern Asia and North America. It lives in fresh and brackish water close to the shore. Originally, the Stickleback was a sea dweller. Only after the last glaciation ten thousand years ago has it populated freshwater. Since then, many phenotypes have developed that have multiple times been described as different species, variations and forms, in part or synonymously (Prud'homme *et al.*, 2006). The differences in body structure among the different forms are often larger than the differences between different genera in other groups of fish. Among the most striking variations in Sticklebacks are the almost 30-fold differences in size and number of the protecting bone plates (instead of scales). The absence or additional presence of particular fins, as well as significant differences in jaw shape, body shape, tooth structure, the defensive spines and in body coloration is also striking.

Most striking are the two forms of Sticklebacks in many North American lakes. One lives in the open deeper water, and the other lives on the bottom in shallow areas. The deep-water form exhibits a pronounced, spined ventral fin, similar to the original sea dweller. The shallow water form lacks both the fin and spine (Prud'homme *et al.*, 2006).

All of these forms are evolutionarily very young. They evolved after the last ice age, and some forms evolved several times independently from each other. In captivity, these phenotypically very different fish can be crossed with each other artificially. Of course, this does not say anything about their species status, but it has an important practical consequence for the scientist. The genetic cross-compatibility allows for the identification of genes involved in the reduction of the pelvic fin. Surprisingly, the remarkable anatomic alterations apparently trace back to only a very small number of genes that function as the developmental control genes (Kingsley, 2009).

The presence or absence of the ventral spine in the Stickleback essentially traces back to the expression of only a single gene. This gene is named "*Pitxl*" and is (as usual) controlled by various enhancers. One particular enhancer is responsible for the expression of the *Pitxl* gene in the ventral fin, the "ventral fin enhancer." In other body parts, *Pitxl* codes for the same protein, but it affects the development of other body structures. Failure of the "ventral fin enhancer" only causes the failure to express the gene in the ventral fin area and thus specifically affects only development of the one spiny ventral fin. The other enhancers remain intact and continue function.

The Stickleback is an exceedingly instructive example for typologically oriented taxonomists. It shows that the reshaping of entire body structures can trace back to a single mutation.

4.5
What is the Relevance of Differences in Traits between Two Species?

Two traits can appear different or perform different functions but nevertheless be encoded by the same gene. What then are two "different" traits if they are based on the same gene? If the structural genes hardly differ between two species, what is it that differs if two species have different traits? If one species possesses a certain trait and another does not, we assume intuitively that the first species has acquired this trait at a later time. However, traits can also be lost. The control of the structural genes via enhancer switches makes apparent that anatomical and morphological alterations during evolution can be based on the repeated switching on and off of structural genes that otherwise do not change during the speciation processes.

What is a newly evolved trait if its coding gene is not new? This consideration questions the concept of Willi Hennig's apomorphy (Hennig, 1966) (Chapter 7). Hennig's cladistic taxonomy defines the origin of two daughter species from an ancestor species (stem species) by the fact the two new daughter species must possess "newly evolved" ("derived") traits (e.g., a color pattern or the form of a beak). Derived traits are called "apomorphies." In modern cladistics, the term "specifier" is used (de Queiroz, 1999). Apomorphic traits are opposed by plesiomorphic traits. Plesiomorphies are phylogenetically older traits that were already present in the ancestors of newly developing daughter branches of the genealogical tree and thus are common to both daughter species and their ancestors. A trait in a daughter branch is therefore either an apomorphy or a plesiomorphy. Only in the latter case does the trait have a species-defining meaning.

However, apomorphies or specifiers are taken as a given, without these traits being broken down in their complexity. What is an apomorphic trait in the Hennig's sense, and what is a specifier in the sense of de Queiroz if nothing has changed except for the point in time or the intensity in switching on or off of an already preexisting and hence plesiomorphic gene? Newly derived apomorphic and evolutionary ancient plesiomorphic traits can be controlled by a single gene even without this having changed.

The genetic status of an apomorphy can be reversed at any time, and the genetic status of an apomorphy can occur several times in independent branches of the taxon's evolution. The results of "evo-devo" research involving mammals, insects and plants show that the development of markedly different morphological structures can be traced back to the reuse of just a few development-controlling "master" genes that do not distinguish themselves from each other in different groups of organisms. Only the modality of expression has changed (Cronk, Bateman and Hawkins, 2002). What then is the previously non-existent evolutionary novelty, the apomorphy in the sense of Hennig? What is a "newly-evolved" trait?

What is the trait itself in its final, finished state, as can be observed on the body of an animal? Only in the rarest cases is it a simple component, such as a molecule for a pigment, a blood group or a muscle protein. In most cases, a trait is complex and composed of an abundance of individual traits. These individual partial components can be either apomorphic or plesiomorphic, which results in a trait that is both apomorphic and plesiomorphic. In comparing two traits, certain partial components of these traits can be homologous to each other and have a common ancestor, but additional components of the same trait may be non-homologous to each other. Two complex traits that are compared to each other are often neither homologous nor are they non-homologous (see below).

What are hair length and hair density, facial proportions or body size? If humans and chimpanzees have different hair density and facial proportions, what trait is actually meant then? It may very well be the case that from the abundance of control mechanisms influencing facial proportions only a single component accounts for the differences between humans and chimpanzees. A newly evolved complex so-called apomorphic trait may well consist of many components that almost all are plesiomorphic, while only one is really apomorphic. What then is an apomorphic trait?

Whether taxonomic traits are truly similar to each other cannot be decided by a superficial examination. We are too susceptible to optical deceptions and the differences between a spontaneous overall impression and the underlying details are too large if a trait's components are examined with more refined methods. Then, significant differences that were not visible before can come to light. Obviously, a bird's wing and a bat's wing are similar in appearance as well as function. However, which component should be considered when talking about resemblance? It should be disquieting that this cannot be expressed in numbers, that is, quantified. When is the extent of similarities greater than the extent of dissimilarities? Could not a perception of resemblance based on a spontaneous examination be examined with a greater scientific foundation?

If two species are distinguished by their traits, this is only possible if the traits are assessed. However, the fatal flaw is that there are neither quantitative nor qualitative rules for such an assessment:

1) **Quantitative**: The number of trait differences approaches infinity, and trait differences do not only distinguish two species but also distinguish two individuals of the same species. There is no possibility for determining a limit; that is to say, a given value of trait differences, whose exceedance would mean that the level of species has been reached (Dupre, 1999). How similar do two organisms have to be for them to be similar enough to belong to one and the same species (Hull, 1997)? Do there have to be ten or a hundred or more trait differences to justify the status of species? If it has already been determined that there have to be ten traits, what is then one of these traits? Is it body hair of a certain length and color, for example or does the body hair already contain ten traits because the hair consists of ten components or is controlled by ten genes?
2) **Qualitative**: There is also no possibility of defining categories of quality as generally mandatory, which have to be complied with by particular traits to reach taxonomic relevance, while those traits that do not reach these quality requirements would then be taxonomically useless. Every attempt to assess traits can only be based on a highly subjective decision. For example, a human and a rhesus monkey can have exactly the same blood group. However, two humans in comparison to each other or two Rhesus Monkeys in comparison to each other might have markedly different blood groups. Why do all Rhesus Monkeys belong to one and the same group, just because of a trait definition? Why do humans and Rhesus Monkeys belong to different groups just because of a trait definition? The underlying problem is that it concerns unerring intuition. It has often been said that species can only be determined by the covariation of multiple traits and not by traits being considered separately. Here, it is about the question of whether it is possible to have a fundamental law in taxonomy from which it might be inferred that blood group cannot be used as a taxonomic trait of delimitation. What law says which traits are allowed to be used and which traits are forbidden?

4.6 Traits that are Used by the Species to Distinguish Themselves

If species live in the same region, they can only remain in existence for a long time if their members can recognize and distinguish each other. If they cannot, they would merge with each other and cease to be separate entities. Therefore, all biparental organisms that live in the same region have certain identifying traits to make the correct partner choice. Females are especially picky. In most animal species, it is their part to decide which particular individual is permitted sexual contact. Sexual willingness of the female does not arise if the male does not display the correct optical, acoustical or olfactory signals. Only if these traits "fit" do the females show the necessary disposition required for reproduction. This biological phenomenon is called "female choice."

Signals such as these can be particular color traits. The female recognizes these signals with a high degree of accuracy and distinguishes the conspecific male sexual partner from foreign partners using these traits. Duck males, peacocks and birds of paradise have especially impressive color signals that distinguish the species from each other. In addition to optical partner identification signals, there are also acoustical ones. Many bird songs are species-specific. They are presented by the males and serve for the females to distinguish their own species from a foreign one. Olfactory signals for species recognition are also widely distributed in the animal kingdom, for example, in Clearwing Moths (*Sesiidae*) and the representatives of some moth families. There are examples in which a single signal molecule sent by the female is enough to unerringly attract the male for copulation.

Above, the use of traits for species classification was called into question because traits are subjectively selected by humans for their own intentions for differentiation. However, in the cases of partner recognition signaling, only traits play a role for which the animals can recognize themselves as sexual partners. This seems to be a way out of the dilemma because with partner recognition signals the species protect themselves from false mating. The selection of those traits by which the members of a species recognize each other for a taxonomic delimitation of species is without doubt tempting. The exclusive application of only these traits for taxonomic classification appears to be reliable. This approach seems to exclude the traits that have nothing to do with species specificity.

Remarkably, one of the fathers of the species concept of the reproductive community, Theodosius Dobzhansky, was already aware that there have to be two different kinds of traits distinguished: those that are responsible for partner choice and those that have nothing to do with partner recognition (Dobzhansky, 1937). "The genetic factors responsible for the production of the isolating mechanisms appear to constitute rather a class of particular traits by themselves." He introduced the term "isolating mechanisms" and defined them as an isolating traits or "any trait that hinders the interbreeding of groups of individuals." Those traits would "make interbreeding (with non-conspecifics) difficult or impossible."

Unfortunately, this assessment cannot be generalized. Several species live in different geographical regions and overlap each other only within limited districts. If, however, two species live in different geographical regions, in these regions, the species-specific recognition signals are unnecessary. There is no selection pressure for species distinction and, hence, for the development of species-specific recognition signals. Only in the overlapping regions where two species live sympatrically and meet each other do they have to be able to distinguish each other in a species-specific manner. In all of the other regions, species-specific recognition signals are unnecessary.

As a consequence, some species carry species-specific recognition signals only in the overlap region and lack these signals where they do not meet each other. A well-documented example for this is the European species pair the Pied Flycatcher (*Ficedula hypoleuca*) and Collared Flycatcher (*Ficedula albicollis*) (Saetre *et al.*, 1997). The Pied Flycatcher is widely distributed throughout central and northern Europe to western Siberia, while the Collared Flycatcher is predominantly restricted to the southeast of Europe.

Because of the great resemblance of both species, there is a problem of correct partner choice, which relatively often results in species hybridizations (Chapter 6). In the Flycatcher example, it becomes apparent that the species identification traits used by us humans are not the same as those traits that birds recognize and distinguish themselves. The recognition trait relevant to us humans is the eponymous white collar of the Collared Flycatcher males that the Pied Flycatcher lacks. However, the birds obviously cannot use this trait for species discrimination. The female flycatchers use another trait for recognizing the correct sexual partner. Pied Flycatcher males are pale brown colored on the top and distinguishable from the Collared Flycatcher males, which are colored deep black on the dorsal side. Experiments have shown that this plumage coloration trait is indeed used as "species recognition trait" by the Pied Flycatcher females to correctly recognize their males and to distinguish them from the Collared Flycatcher (Saetre *et al.*, 1997).

Interestingly, in the vast regions of Western and Northern Europe where the Pied Flycatchers occur alone, the gray plumage coloration does not occur. There the males of the Pied Flycatcher are colored pitch-black similar to the Collared Flycatchers and are accepted by female Pied Flycatchers because confusion with the wrong species is impossible. In these regions, the Pied Flycatcher females do not need to distinguish their males from those of the Collared Flycatcher. Consequently, the species recognition trait of the pale brown back is missing there. The species recognition trait of a pale brown back as opposed to a black back in the case of the Pied Flycatcher has apparently developed under selective pressure to facilitate species recognition for females only in the regions where overlap occurs.

A similar example of this phenomenon occurs in the species pair of the Western Rock Nuthatch (*Sitta neumayer*) and Eastern Rock Nuthatch (*Sitta tephronota*) (Vaurie, 1951). The Western Rock Nuthatch occurs in southeastern Europe and southwestern Asia, the Eastern Rock Nuthatch lives more to the east in Iran and further east into Pakistan. Both species overlap in Iran, and only there do the two species differ distinctly in some traits. In the regions of exclusive occurrence, the two species are nearly indistinguishable. However, in the overlap region, the Eastern Rock Nuthatch is larger than the Western Rock Nuthatch and most importantly has a larger and longer beak. It is likely that the difference in body size and beak form in the region of joint occurrence of the two species prevents their merging because it serves both Nuthatch species as a species recognition trait for a correct partner choice. In addition, the difference in beak size in the overlap region seems to lead to different food preferences, thus avoiding competition. This does not make sense in places where both species do not occur jointly and thus do not have to compete with each other.

Another example for the phenomenon that species-specific traits are only realized in the region of overlap of two species is provided by members of the genus *Zygaena* (Burnets) (see Color Plate 4). The three species of the *Zygaena-transalpina* group, *Zygaena transalpina*, *Z. hippocrepidis* and *Z. angelicae* differ only in the geographical regions of joint occurrence where they encounter each other (Ebert *et al.*, 1994). Only in these small overlap regions do they show their respective otherness and differ in their choice of habitat as well as larval food plants and respective seasonality for when the imagines hatch. For example, in the overlap region, *Z. angelicae* occupies beech groves, while *Z. hippocrepidis* prefers dry grassland. *Z. angelicae* prefers the Scorpion Vetch (*Coronilla coronata*) as a larval food plant in the overlap region, while *Z. hippocrepidis* feeds on the Horseshoe Vetch (*Hippocrepis comosa*). *Z. angelicae* flies from June to the middle of July in the overlap region, whereas *Z. hippocrepidis* is there from the middle of July to the middle of August (Hille, 1995). Through these evasion strategies, competition and also sexual mixture of the different species is avoided. In the regions in which the three species occur alone and in which a selective pressure for distinction and evasion does not exist, these species-specific differences are not realized. In these regions, it is nearly impossible to distinguish the three species.

It is often argued that this case is an example of incipient speciation, which only has its "real" beginning where the species encounter each other. Only there could one speak of speciation (Ebert *et al.*, 1994). The populations in the regions where the three species do not occur jointly had not achieved the status of "real" species. However, no argument exists that posits that the current status of the *Zygaena-transalpina* group has to be transitory.

Indeed, there are examples in the animal kingdom that favor the contrary view that species differences can disappear in geographical regions where they do not make sense anymore. For example, the Mallard (*Anas platyrhynchos*) and the Pintail (*Anas acuta*) are closely related species. They live sympatrically in major parts of the Holarctic. There, the females of both species have to be able to distinguish their males from the respective alien species. The plumage coloration of the male, which is distinctly different in both species during the mating season, plays a role in the partner recognition of ducks.

However, the necessity for correct partner recognition ceases to exist in places where only one of each species occurs, and thus the discrimination of species is unnecessary. This is the case with the Mallard in Hawaii and the Pintail in the Kerguelen Islands, a group of islands in the southern Indian Ocean. Most of the male Mallards in Hawaii and most of the male Pintails in the Kerguelen Islands are, like the female, uniformly brown-colored for the whole year, even during the mating season. Thus, the different species become similar. They do not wear the species-specific mating plumage, probably because they do not have to contrast with the respective related species.

In this case, one cannot argue that the similarity in traits between two species is a sign for incipient speciation that has not been completed. The opposite is true. This is a case of evolutionary retrogression of species recognition traits because the selective pressure to maintain them has subsided. The species recognition traits seem to have become lost because it was no longer required to maintain them. The Mallard in

Hawaii is designated as the subspecies *wyvilliana*, and the Pintail in the Kerguelen Islands as subspecies *eatoni* (del Hoyo, Elliott, and Sargatal, 1992).

These examples make clear that in some species the species identification traits are only of local importance. They are not valid for all organisms of the species. Such traits are useless as recognition traits for the species as a whole.

4.7
A Species cannot be Defined by Traits

As explained in other parts of this book, a species cannot be defined by its traits for many reasons. It is hard to actually define what a trait is. The number of traits is almost infinite, and a choice has to be made as to which traits permit acceptance of a species as a species; however, every choice is subjective. No single species recognition trait is essential for the species because no trait is sufficient and necessary to define an organism as belonging to a species (Chapter 3). The membership of an organism in one species can only be determined by the covariation of several traits.

However, if a species cannot be defined by even a single essential trait, then the characterization of a species by a catalog of covarying traits is at best an operational definition. An operational definition does not tell us anything about what a species is, but only how to distinguish a species from another species (Mahner and Bunge, 1997). It only tells us which properties a species has (Chapter 2). However, trait recognition is different from defining criteria for a species. Traits are symptoms of a species, so to speak (de Queiroz, 1999). If one's limbs are aching, this is a symptom of the flu. However, this limb pain does not define what the flu is.

It is tempting to equate similarity in traits with species membership, because in several cases members of the same species indeed resemble each other in traits. There are biological reasons why organisms of common descent or the members of a gene-flow community have similar traits.

Ancestors pass on their traits to descendants, and because of this, the descendants resemble their parents and grandparents. Because of kinship, the descendants (the siblings) also resemble each other. However, this resemblance continuously declines over the course of evolution. The sibling branches continue to lose resemblance over time due to mutations and adaptations to local conditions. They all still belong to same descent community, although their trait similarity may have considerably declined over the course of time.

Also the biological principles behind the species as a gene-flow community explain why the traits of members of a gene-flow community resemble each other. Mutual sexual contact among the organisms generates trait similarity among the organisms. To a limited extent, biparental sexual recombination of even distant organisms causes a homogenization of traits, even between individuals that previously diverged in their trait resemblance. Many species have trait similarity because their organisms continually mix through genetic recombination and are thus kept homogeneous (see Section "Why do the individuals of a species resemble each other?" in Chapter 6). However, if the geographical distance between the members of a species exceeds a certain limit, homogenization of traits by genetic recombination declines, and local races arise (Chapter 5).

Trait similarity and trait differences are certainly phenomena that are causally connected to biological rules. Similarity of traits could be the result of both common descent (the species concept of kinship) and sexual gene exchange (the species concept of the reproductive community). If species are grouped by kinship or as gene flow communities, similarity in traits has a certain probability. However, the trait similarities among the members of a descent community, as well as the trait similarities among the members of a gene-flow community, can be substantially weakened. Furthermore, the similarities, or rather the dissimilarities, of traits can also have other causes (see several of the Color Plates).

Thus, it should not be a principle to treat the members of a species as equivalent with regard to their traits because often, they are not equivalent in this manner. The inference of status as a species from trait equivalence has the merit only of an initial suspicion (Mahner and Bunge, 1997). In discovering two groups of organisms with different traits, one can only claim a lower or higher probability for the assumption that these groups are two different species. Trait similarity in general is not a criterion of species membership. It is not possible to establish a species definition based on criteria that apply only by the majority of examples, but not by all.

Since the organisms of different species are distinguished from each other by their traits, but the organisms within a species are also distinguished by their traits, the question arises of whether intraspecific differences (that is, differences between organisms of the same species) are based on trait qualities that are different in type from interspecific differences (that is, differences between the organisms of different species). Basically, there are three lines of thought on this issue:

1) A species has fewer intraspecific differences than there are interspecific differences between related species.

   This relationship could be true, but it does not have to be true in all cases. With respect to DNA sequence traits, which have the advantage, compared to morphological traits, that they can be counted, individuals of the same species could differ more from each other than the members of different species (Avise, Walker, and Johns, 1998; Verheyen *et al.*, 2003). Thus, it is not admissible, at least in the case of evolutionarily young species, to elevate the extent of trait differences to a species criterion, in the sense that small differences indicate the same species, whereas more pronounced differences must indicate a different species (see the criticism of barcoding below).

2) The differences between species rely on special trait qualities, whereas intraspecific differences have different trait qualities.

   Is there any type of trait that marks a species as a species, as in chemistry, where the number of protons determines the identity of a specific element? Can we simply choose a morphological trait (e.g., the number of limbs) and divide the animal kingdom accordingly into groups in a law-like manner, similar to a chemist compiling a table of elements? This approach has certainly been attempted in the past in taxonomy research (Chen, 2002). The modern response, however, is no. Nothing distinguishes traits that are responsible for intraspecific differences from traits that are responsible for interspecific differences. After all, it was this realization, among other concerns, that induced Darwin not to distinguish between variations among species and those between groups of individuals within the same species (Darwin, 1859).

Admittedly, in some peculiar cases there are examples where a simple trait quality can be applied for taxonomic classification, for example, the number of legs. For example, the tetrapods (four-footed animals, all of which are land-living vertebrates, that is, classes of amphibians, reptiles, birds and mammals) and hexapods (six-footed animals, i.e., insects) are indeed classified by their number of legs. But these examples refer to very specific groups. The number of legs is not a universal classification criterion for all organisms. Can we take specific segments of the genome, for example, sequences for the regulation of structural genes, to determine the taxa accordingly? The answer is no; this approach does not work, either.

3) The trait differences between species separate the organisms into distinct groups, whereas intraspecific trait differences are distributed equally among the individual organisms of a species.

   This approach is also not valid, because of the diverse examples of intraspecific polymorphisms, as well as examples of geographical races, that divide the organisms of one species into distinct groups. Many intraspecific differences are not homogeneously distributed among the organisms of a species. The trait qualities that characterize morphs or races divide the organisms within a species just a swell into distinct groups as species are divided into groups.

4.8 What are Homologous Traits?

When are two traits homologous to each other? Two traits can look markedly different or perform markedly different functions but, nevertheless, can be encoded by the same gene. One gene could be expressed as markedly different phenotypic traits in different organisms. What, then, are two "different" traits, if they are based on the same gene? What are homologous traits, if they are based on the same gene?

Homology is a biological concept that is difficult to understand and is therefore often understood incorrectly. The source of the difficulty of understanding the concept of homology lies in the fact that, in evolution, there are two markedly different ways for phenotypic traits to resemble each other. Trait similarity can result from common descent but can also result from similar selection pressure. These sources for trait similarity are unrelated but produce the same final result.

The key criterion for the definition of homology is common descent (Mindell and Meyer, 2001). If two cells emerge from a single cell through cleavage, then they are homologous to each other, even if they take on markedly different tasks and attain a markedly different appearance. If two DNA sequences emerge from a single sequence, then they are homologous to each other, even if they become different through mutation and take on different tasks.

The four-footedness of a newt and a horse traces back to a common ancestor. This may be considered an example of homology because it is unrelated to environmental adaptation. The newt, with its clumsiness, would perhaps move better on six feet. On the other hand, the fins of a seal, a whale and a fish look similar; however, this type of construct obviously does not constitute homology, because it is a result of environmental adaptation. The striking similarity of the fins has nothing to do with common descent but is a result of the same selection pressure. This phenomenon is called convergence (parallel evolution). When similar traits are observed in different organisms, it is often difficult to substantiate whether the traits are similar because of homology or because of convergence.

For a long time, it has been clear that the many similarities between Swifts (family *Apodidae*) and Swallows (family *Hirundinidae*) trace back to convergence. Both families belong to different orders; they are, therefore, only slightly related to each other. The streamlined body and the sickle-shaped wings are not the same between Swifts and Swallows because they share the same ancestor but because they share the same behavior. Only a few years ago, it has been detected that the great similarities in morphology and behavior between New World Vultures (*Cathartidae*) and Old World Vultures (*Aegypiinae*) also do not result from common descent but instead result from convergence (del Hoyo, Elliott, and Sargatal, 1994). The seven species of New World Vultures, to which the Andean Condor (*Vultur gryphus*) belongs, are more closely related to Storks (order *Ciconiiformes*) than to the Raptors (order *Falconiformes*). The short, raptor-like hooked beak and the bald facial skin are adaptations to the gathering of food from the interior of carcasses; these are not traits that trace back to a common descent.

At first view, one would say that New World Vultures and Old World Vultures are not homologous to one another. But the question arises whether the term homologues can be applied to an entire organism, such as a Vulture. The concept of homology appears to refer to replicating structures, such as cells or DNA molecules. Can a trait, in principle, be homologous to another trait? Can phenotypic traits such as body parts, construction plans, behavioral patterns, and developmental paths in principle all be treated as homologous to each other? Two phenotypic traits cannot have a common ancestor because they do not replicate. All morphological structures are complex, meaning that they are assembled from individual parts. It is not possible to trace these structures back to homologous genes. If one tries to trace complex structures back to common ancestors, this succeeds only if the complexity of the morphological structure is dismantled into its parts. Then some of the parts are homologous to each other and others are not.

Homologous genes in no way need to encode similar phenotypic traits. The "evo-devo" research of recent years has shown that many single genes can change their role in the developmental paths of different species, so that they then influence markedly different morphological structures. Examples of this phenomenon are the transcription factors *distal-less*, *engrailed* and *orthodenticle*, which, as orthologs in various metazoic taxa, influence the pattern formation of distinctly different structures (Mindell and Meyer, 2001). Thus, markedly different developmental processes can be governed by homologous genes.

In contrast, the same or similar developmental processes in different taxa can be controlled by markedly different genes that are non-homologous to each other. For example, gastrulation in the early embryonic development is referred to as homologous across all vertebrates, although this developmental process is controlled by significantly different genetic mechanisms. Thus, it is not consistent to label the gastrulae of different animal phyla as homologous. This constraint is at best present for some of the partial components that are involved in gastrulation. Only these components can be considered to be homologous. Another example are the supposedly homologous body segments of *Drosophila* and of the Migratory Locust (*Locusta*), which are, however, genetically controlled by different transcription factors. Thus, these body segments, as a whole, cannot be homologous to each other, because only some components are homologous. Similarly, one would not label two things as being colored red when they are colored both red and blue, so that the commonness of both is not only the color red.

Furthermore, the concept of homology becomes less and less useful in its significance and meaning when going further back into phylogeny. If two structures descend from a common ancestor that traces far back into phylogeny, then the assessment of the common descent and, thus, the prevailing homology relation does not produce any specific gain in scientific knowledge, although it is logically unassailable to call all of these structures homologous. Because of the common descent of all of life, every structure is, in the end, homologous to every other structure.

4.9
The Vertebrate Eye and the Squid Eye: They Cannot be Homologous Nor can they be Non-Homologous

Most traits are complex. This observation means that what is considered by the taxonomist to be a trait in reality is a complex structure that is composed of several components, because evolution has structured the traits of a phenotype according to a modular principle. The individual components of a single trait sometimes share very different evolutionary histories. This scenario clearly means that many traits cannot be homologous and also not non-homologous, because they are composed of components with different phylogenetic descents. Even several proteins can be partially homologous to each other and partially not homologous, because different proteins can be modularly assembled from different domains. If two proteins in two different organisms are compared, some of the domains of a specific protein might be homologous, whereas other domains of the same protein might be convergent parts that have independent evolutionary origins. Examples of this phenomenon are some trans-membrane domains and specific receptor domains (Ast, 2005).

The concept of homology is often misunderstood. The disagreement of scientists on the concept's formation has led to the "homology problem" (Bock, 1989). Researchers with differing interests, experiences and objectives have defined the concept of homology differently, and the definition has also changed over time (Mindell and Meyer, 2001).

Here is an example of the difficulty of concept formation: The relationship between the vertebrate eye and the squid eye is a popular textbook example of convergence. Both of these visual organs have evolved independently from each other over the course of evolution; the vertebrate eye evolved from the brain, and the squid eye evolved from the skin. In spite of these different origins, both eyes show amazing similarities in the entire organization of the lens, vitreous body and retina. These similarities are non-related secondary similarities that are caused by equivalent selection pressures. For this reason, the vertebrate eye and the squid eye are considered to be convergent.

However, in spite of this scenario, there are also reasons to consider them to be homologous. This proposal is not contradictory; instead, it is simply based on the fact that both organs are assembled from parts that have similarities that are based on homology, which is a real kinship, as well as from parts that have similarities that are based on parallel evolution. Some components of the vertebrate eye and the squid eye are homologous, whereas other parts are not homologous and, instead, are "only" convergent.

For example, the gene *pax* 6 (paired-box gene) is an important key gene for the eye's development and function. *Pax* 6 is a tissue-specific transcription factor that regulates the eye's development in an early development stage and plays a role with regard to the differentiation of the lens and retina (Gehring, 2002). The *pax* gene is strongly conserved and, in vertebrates and squids, originates from a common ancestral stem gene. This gene is, therefore, a distinctly homologous component of both eyes – the squid eye and the eye of the vertebrate (Piatigorsky and Kozmik, 2004). Other genes, however, which also shape the traits of the eye, are not related with respect to kinship. Thus, the eye is composed of traits that are partly the products of homologous genes but are also partly the products of non-homologous genes.

What, then, is a homologous organ? Is homology based on the construction plan (in which case the vertebrate and squid eye would not be homologous to each other), or is homology based on specific traits (in which case the vertebrate and squid eye would be homologous as well as non-homologous to each other)? Clearly, it is not possible to apply the concept of homology to complex structures in biology. Organs are almost never structures of common descent. Their components and tissues are not descendants of a common ancestor. Most organs are assembled from individual parts, each with a different evolutionary history. Some of these components could be homologous but others of the same trait could well be non-homologous. If a complex trait is compared between two organisms, it often can be neither homologous nor non-homologous. Only the individual parts can be homologous to one other. One would not consider two objects as being colored red when they are colored both red and blue.

It appears that the basic question of whether two organs are homologous or not was a meaningless question from the start. The application of the concept of homology to entire systems is an incorrect way of reasoning, because this approach ignores the modular principle of evolution. It may be that an aversion of conservative biologists against reductionism in biological science has supported the view for the "whole." Entire organs are considered to be more than the sum of their parts, and this world view may have favored the aim to conceive organs and organisms as whole entities that should not compartmentalized.

4.10 The DNA Barcoding Approach – is Taxonomy Nothing more than Phylogenetic Distance?

The technique of barcoding refers to the identification of short DNA sequence sections, mainly a 648-bp region of the cytochrome c oxidase I gene (COI) of mitochondria, with the intent of determining taxonomic classifications. These sequences are compared among the organisms. From the extent of sequence differences, a conclusion as to the species identification of the organisms is drawn. Undoubtedly, the differences in DNA sequences indicate the differences in the phylogenetic distances among the respective organisms. However, the genetic distances between groups of organisms are not the same phenomenon as the species differences (Chapter 6). Many studies have documented that reproductively well-separated species hardly differ genetically, and in turn, there exist species with large intraspecific genetic heterogeneity. Hence, barcoding does not lead to the same entities as other species concepts. Because the barcode determines only the genetic distance among groups of organisms, it inevitably comprises all of the evolutionarily young species as a single common species, and it subdivides many old species into separate species.

The method of DNA barcoding has been highly praised. It is advertised as a "renaissance in taxonomy" (Miller, 2007) or as "taxonomy of the twenty first century" (Steinke and Brede, 2006). Its application to butterflies (Hebert *et al.*, 2004), birds (Hebert *et al.*, 2004) (www.plosbiology.org) and many other organisms led to the discovery of many supposedly new species. One of the first publications of Hebert and colleagues (Hebert *et al.*, 2004) on the Central American Hesperid butterfly *Astraptes fulgerator* has the promising title "Ten species in one: DNA barcoding reveals cryptic species in the neotropical skipper butterfly *Astraptes fulgerator.*"

However, the cogent question is whether all of these promises and visions can withstand a stanch reconsideration of the biological units that are truly identified by the barcoding method. One year before the barcoding method was introduced, J. Willem H. Ferguson from the University of Pretoria published a paper in the Biological Journal of the Linnean Society that was entitled "On the use of genetic divergence for identifying species" (Ferguson, 2002). The abstract of this article begins with the following three sentences:

> "Degree of genetic divergence is frequently used to infer that two populations belong to separate species, or that several populations belong to a single species. I explore the logical framework of this approach, including the following assumptions: (i) speciation takes place over very long periods of

time; (ii) reproductive isolation is based on the slow accumulation of many genetic differences throughout the genome; (iii) genetic divergence automatically leads to reproductive isolation between species; and (iv) premating and postmating reproductive isolation have a similar genetic basis. I argue that so many exceptions to these assumptions have been demonstrated that they cannot be used with any reliability to distinguish different species."

All of these four assumptions are not generally valid and are arguments against the biological foundation of the barcoding approach. If you believe in the truth of all of these four assumptions, then you may believe that barcoding is the real alternative to all of the conservative species concepts. However, the four assumptions do not hold true for several examples of species formation.

What is the basic principle of the barcoding approach? All DNA sequences are subject to mutation. Many mutations disappear soon after their appearance because they are eliminated by selection or genetic drift (Chapter 5). However, a few mutations survive over the course of time. If the branches of the phylogenetic tree become separated from each other, then different mutations accumulate in the different lineages in the course of longer evolutionary timeframes. The lineages diverge, and the DNA of the genomes becomes more and more different among the organisms. If it is viewed very roughly, this mechanism proceeds proportionately with time. For this reason, it is also referred to as a "molecular clock." It is possible to tell, from the number of base differences between two DNA sequences that are homologous to each other, how long ago the time was at which two lineages separated from each other. DNA sequence differences are a measure of the kinship of two lineages. Many base exchanges means that there is a distant kinship and few base exchanges means that there is a close kinship.

This principle is used by the method of "barcoding." It was in 2003 that Paul Hebert from the University of Guelph, Canada, declared "DNA barcoding" to be the trendsetting method for identifying species (Hebert, Ratnasingham and deWaard, 2003). In this way, the pursued objective is easily understood. Barcoding means the sequencing of a 648-bp region of the cytochrome c oxidase I gene (COI) from the mitochondria and its comparison among different organisms. Because this sequence can be read routinely by automatic scanners in the same way as a supermarket scanner distinguishes products using the black stripes on the packaged goods, this technique has been called "barcoding."

The 648-bp region is being used as the standard barcode for almost all animal groups. It has been shown that 95% of the tested species of various animal groups possess distinctive COI sequences that allow species identification. Mitochondrial DNA sequences do not recombine their genomes after sexual merging of the sperm and egg, and therefore, they have some advantages compared to nuclear DNA (Ballard and Dean, 2001). In plants, however, the COI gene of the mitochondria is not an effective barcode region, but two gene regions in the chloroplast have been found to be suitable for species identification in plants. It basically remains unexplained why a specific mitochondrial gene region should work for almost all animal species but not for plant species. Mitochondria play the same biological role in animals and plants. How is it possible that a mitochondrial DNA sequence is suited for species definition in animals, but not in plants?

Of course, the measurements of differences in DNA sequences allow scientists to draw reliable hypotheses about the point of time of the phylogenetic separation of two organisms. However, the advocates of barcoding go a significant step further, thereby opening themselves to vulnerability. They declare two organisms that exhibit a significant sequence difference in their DNA to be different species. The underlying logic is simple: If two organisms contain two sequences that have separated from each other umpteen years ago, then these two organisms must be different species.

Although barcoding is strongly scientific, because it is free of subjective evaluations, the classification of organisms by their barcode provides only information about the genealogical proximity or distance of certain organisms and not more. But when are two groups of organisms distant enough from each other to be separate species? The method of barcoding does not provide information about this concern, because it does not imply any criteria for the demarcation of one species from another.

It is possible to declare the phylogenetic distance to be a species concept. This suggestion, however, is a different species concept than other concepts. Most publications describing new species on the basis of barcoding (e.g., Hebert *et al.*, 2004) make the mistake of not expressly pointing out that they use a very one-sided species concept. Thus, they make themselves unnecessarily open to attacks.

The species concept represented by the barcoding method ignores the fact that there are young and old species. Phylogenetically young species are species that originated only a short time ago; therefore, it is genetically difficult to distinguish them from each other. Old species are species that have existed for millions of years. Young species consist of organisms that are genetically similar. Old species consist of organisms that are genetically different from each other, unless they passed through a genetic bottleneck in recent times. The passage through a bottleneck reduces their genetic diversity. If the species did not pass through a bottleneck, however, old species achieve a high level of intraspecific genetic diversity, especially if the individual populations live geographically distant from each other (Garcia-Ramos and Kirkpatrick, 1997; Avise, Walker, and Johns, 1998; Varga and Schmitt, 2008; Habel *et al.*, 2009).

Phylogenetic distance is different from the concept of reproductive incompatibility (Ferguson, 2002). In some cases, a distinctive reproductive isolation exists between two species, although they are genetically hardly divergent (Schliewen *et al.*, 2001). In other cases, no reproductive isolation exists between two specific populations that diverge very far from each other genetically (Coyne and Orr, 1997). The concept of barcoding must split them up into different species, although they belong to only one species according to the species concept of the gene-flow community. Anyone who turns that concept into a species, defining a species based on being beyond a minimum of phylogenetic distance, must be aware that, in doing so, he does not represent the species concept of the gene-flow community. Barcoding provides data for the distinction of close or distant kinship in the comparison of two populations but does not make any statements regarding their reproductive compatibility.

The many species of Cichlid fishes in Lake Victoria provide a strong example of distinct reproductive isolation between species that have barely genetically diverged

(Stiassny and Meyer, 1999; Verheyen *et al.*, 2003) (Chapter 6). Over presumably less than ten thousand years, there has been an adaptive radiation of enormous extent. Lake Victoria was completely dried out approximately ten thousand years ago. By some mechanism, a few Cichlids entered the lake afterwards. In the short time until the present, 400–800 species originated from this very small founder population.

These species drastically differ from each other with regard to their traits, which are responsible for ecological adaptation and partner recognition. They have varying body sizes and color patterns, varying fin forms and also widely varying mouth and jaw shapes, which reflects an adaptation to varying types of food consumption and mutual partner recognition. The Cichlid species of Lake Victoria and other African lakes differ genetically even less than humans of different populations (Schliewen *et al.*, 2001; Verheyen *et al.*, 2003). This example clearly documents that the phylogenetic distance between two species is not a measure of the extent of their mutual reproductive incompatibility, and it is clearly not a measure for what is thought to define a species in a general sense.

The equation of phylogenetic distance and reproductive isolation results from a misconception of the extent of gene flow within a species (Chapter 6). The exchange of alleles among distant populations of a species could become very weak across long geographic distances. In many cases, newly arisen allelic mutants do not reach all distant populations within a species and, thus, they promote genetic divergence within a species. Alleles are not "fixed" anymore along the entire distribution area of a gene-flow community, but only within limited ranges of the entire distribution area. As a consequence, geographically distant populations could differ genetically, although they still belong to a connected gene-flow community (Ehrlich and Raven, 1969). A scenario in which a gene-flow community attains considerable adaptive intraspecific genome variability is conceivable (Andolfatto, 2001), and a gene-flow community could be composed of several local races that are genetically distinct (Chapter 5).

The extent of internal genetic cohesion of the individuals of a biparental species has been overestimated in the past (Mishler and Donoghue, 1994). This belief can be traced back to the advocates of the species concept of the gene-flow community, Theodosius Dobzhansky and Ernst Mayr, who were convinced that gene flow and genetic recombination would be the main forces that make the individuals of a species look so similar to one another (Dobzhansky, 1937; Mayr, 1942). However, Ehrlich and Raven (1969) countered that the extent of gene flow is very limited between geographically distant organisms of many species and suggested that the "effective population size in plants is to be measured in meters and not in kilometers" (Chapter 6). The individuals of populations that are separated by several kilometers may rarely, if ever, exchange genes and, as such, could evolve independently. Their phenotypic similarity could be conserved mainly by selection pressure, not so much by genetic recombination, as Dobzhansky and Mayr assumed (Bradshaw, 1972). Lande (1980) has stressed that there has been an overemphasis on the genetic cohesion of widespread species and argued that "of the major forces conserving phenotypic uniformity in time and space, stabilizing selection is by far the most powerful."

Only a few genes are adapted to the local environmental conditions; therefore, the phenotypic uniformity of the traits of the individuals within a species is not based on genome uniformity. Instead, only a tiny fraction of the genome gives a species its species-specific appearance (Coyne and Orr, 1999). Very few genes are involved in the adaptation of a species to its ecological environment and very few genes are involved in correct partner choice. Only the few genes for local adaptation and partner choice constitute the similarity of the individuals within a species. The primary difference between species is not due to the overall genetic difference that arises very slowly and accidentally over the course of long isolation times. The majority of the genome is not involved in maintaining the identity of a species. However, this portion of the genome contains precisely those sequences that are used as a barcode. These sequences are the ones that are subject to the "molecular clock," and they become increasingly different over the course of time. Differences with regard to the barcode do not provide information about a species-specific adaptation to a specific ecosystem, which maintains the distinctive uniform appearance of the individuals of a species.

The species concept of the barcoders is based on some outdated assumptions that date back to the beliefs of the twentieth century. This concept is based on the belief that speciation always occurs under the conditions of geographic separation and, therefore, usually requires a long time, during which the genome would have sufficient time to diverge. Recent discoveries on speciation processes, however, reveal more and more examples of sympatric speciation, and sympatric speciation has been shown to be a relatively fast process in several cases (Dieckmann *et al.*, 2005). Sympatric speciation refers to sexual selection and/or adaptive habitat niching that separates two groups of organisms into different species at the same location. The erection of crossing barriers in the case of sympatric speciation is subject to a high selective pressure to prevent the remixing of the separating populations. Sympatric speciation is the opposite of the allopatric paradigm of speciation because in sympatric speciation, selection is the driving force, whereas allopatric speciation occurs by pure coincidence and proceeds contingently without selection pressure as the driving force of speciation. Allopatric speciation is solely the result of genetic drift and, therefore, it is a slow process (Lande, 1980). Sympatric speciation, in contrast, can immediately split a gene pool, an action which can occur in a few generations. After only a few generations, the time of separation of the newly originated species is not sufficient for the evolution of barcode differences between the new species.

Another main criticism against barcoding concerns the confusion of two different ontological intentions. A diagnosis is not the same as a definition (see Section "It is one thing to identify a species but another to define what a species is" in Chapter 2). It is one task to check whether an arbitrarily chosen organism is a member of a specific species but another task to call a newly discovered group of organisms a new species (Sterelny and Griffiths, 1999). The mixing of the criteria used to diagnose an already established species with the aim of defining a new species is a misconception that apparently cannot be eradicated from taxonomy. The ontological status of a real species and the teleological approach, saying by which criteria a species can be recognized (identification traits), are two utterly different intentions with which to approach nature (Mayden, 1997; Sterelny and Griffiths, 1999). Identifying an already known species by its barcode is a method of diagnosis, not a method for deciding whether a group of organisms is a species (de Queiroz, 1999; Hull, 1968). The biologist who identifies a species already presupposes its existence and only pursues the epistemic goal of how to best recognize it. Barcoding may be a useful tool for the identification of species, but it cannot be a tool for the discovery of new species.

The sudden success of barcoding is surprising because nothing about it is scientifically new. The use of DNA for species identification is as old as the technical ability to sequence the DNA. What is new are the broad claims, the pompous promises and the public relations. The advocates of barcoding pronounce that whereas conventional species identification requires experienced and professional taxonomists, barcode species identification can be performed by trained technicians, or even machines, because the short DNA sequences can be read routinely by automatic scanners. Even the task of the field collection of animals, for example, insects, may be organized with an industrial-scale collection program having the capacity to deliver large numbers of specimens for analysis. Current sequencing protocols permit the recovery of barcode sequences in minutes. The use of on-board barcode reference libraries will facilitate fast species identification in the field (Hebert and Gregory, 2005).

In 2004, the Consortium for the Barcode of Life (CBOL) was founded. This consortium now includes more than 120 organizations from 45 nations (www.barcodinglife.org). The declared goal of the consortium is to build, over the next 20 years, a barcode library for all eukaryotic life. It is expected that the compiled registration of all species of birds and fishes will require some 0.5 million barcode records. For the estimated 10 million animal species of the entire animal and plant kingdom, a comprehensive barcode library will contain approximately 100 million records. This complete barcode library will be constructed within less than two decades.

The resistance against the many assurances of this method has predominantly grown in the circles of the more classically oriented taxonomists (Ebach and Holdrege, 2005). However, also molecularly oriented taxonomists and biophilosophers argue against barcoding. One of the most destructive critics against the barcoding taxonomy has been challenged by the well-known taxonomists Kipling Will, Brent Mishler and Quentin Wheeler (Will, Mishler, and Wheeler, 2005). They accuse the Herbert school of returning to "an ancient, typological, single-character-system approach." This point of critique is overreacted because it ignores the fundamental difference between DNA traits and phenotypic traits in taxonomic classification. DNA traits directly reflect phylogenetic relationships. In most cases, DNA comparisons are blind to morphological convergence (parallel evolution) and, therefore, cannot mislead a taxonomic classification that intends to be based on kinship (Sibley, 1997). Phenotypic traits, in contrast, are controlled by selective pressure and, therefore, are in many cases the subject of convergent evolution. They often do not reflect kinship. This distinction makes barcoding taxonomy incomparably superior to typological taxonomy.

The four objections of J. W. H. Ferguson, however, introducing this section, are still valid (Ferguson, 2002). They are all arguments against the reliability of the barcoding method in defining what a species is: (1) Speciation does not always require a long period of time; (2) reproductive isolation can exist between populations having only small genetic differences; (3) genetic divergence does not necessarily lead to reproductive isolation between species; and (4) premating and postmating reproductive isolation have a markedly different genetic basis.

5
Diversity within the Species: Polymorphisms and the Polytypic Species

5.1
Preliminary Note

Even today, it is still a common view that organisms that are clearly different with regard to traits are different species and that organisms that are less different with regard to traits are subspecies; this belief system does not withstand a persistent examination. Even on the DNA level, the extent of the differences and the certainty that they indicate the status of species cannot be equated. There are groups of individuals that differ genetically but belong to the same species, and there are groups of individuals that are similar in their DNA sequences but belong to different species. The first are evolutionarily old species, and the second are young species.

Only a tiny fraction of the genome is involved with species differences (Orr, 1991). Species arise and are maintained by the action of very few genes, which have earned the moniker "speciation genes." Speciation genes are genes for which the phenotypic traits cause reproductive isolation (see Section "Speciation genes, pre- and post-zygotic barriers" in Chapter 6). Speciation genes are often responsible for the species-specific choice of food plants or for the occupation of a specific ecological niche. Speciation genes are also responsible for the choice of the correct species-specific sexual partner and hence are responsible for the perpetuation of species barriers. Speciation genes constitute only a small fraction of the genome (Orr, 2001). The vast majority of the genome is not involved in differentiating species and keeping them separate to prevent the merging of diverged populations (Chapter 4).

One should not confuse the species-specific traits that cause reproductive isolation, and thus are responsible for maintaining species barriers, with those species-specific traits that became different by chance, that is, as a consequence of long-term separation leading to the cessation of sexual reproduction and gene exchange. The latter traits characterize the different species only arbitrarily, because they play no necessary role in blocking the gene flow. Evolutionarily young species differ in only a few traits, and these are mainly the traits for the maintenance of the species barriers (Schliewen *et al.*, 2001). Evolutionary old species also differ in those traits that are not involved with keeping species separate from each other but which became different just by chance (Ferguson, 2002).

Do Species Exist? Principles of Taxonomic Classification, First Edition. Werner Kunz.

Trait diversity among the individuals of a species is an effective life insurance for the species.

5.3
Superfluous Taxonomic Terms: Variation, Aberration, Form, Phase, Phenon

Unnecessarily, many terms are used to name intraspecific variants, for example, words such as "variations," "aberrations," "forms," "phases" and "phenons."

Apart from sexual dimorphism as a very drastic example (see Color Plates 5 and 6), there are numerous other examples of substantial differences between members of the same species. Different phenotypes could occur between immature and adult individuals of a species or between seasonal generations: the individuals of certain butterfly species that hatch in spring could look very different from the second generation of the same species that hatches in summer, for example, the Map Butterfly *Araschnia levana*. There are numerous historic examples in which these polymorphic types were classified into very different species, in the past.

Many species of large to medium-sized herons display an impressive example of polymorphic plumage coloration. The members of some heron species can be either white or black. This difference can arise from the developmental age (young birds are white, adult birds are dark) or from a dimorphism (adult birds within the same population can be white or dark). For example, immature birds of the Little Blue Heron (*Egretta caerulea*) are snow-white in the year of their birth, but the mature animals are completely black with a slightly cinnamon-red hue. In five different heron species, the adult birds can be either white or dark. Both morphs live alongside each other in the same region. Among the normal dark gray-colored Great Blue Herons (*Ardea herodias*) in the New World, a completely white morph occurs. The Little Egret (*Egretta garzetta*) in the Old World usually is entirely white. However, in rare cases, there also occur totally black morphs. The Central-American Reddish Egret (*Egretta rufescens*), the African Western Reef Heron (*Egretta gularis*) and the Australian Eastern Reef Egret (*Egretta sacra*) can either be white or entirely black. Both morphs of these three species occur alongside each other and mate with each other, apparently without an assortative preference for white or black.

Similarly, Snow Geese (*Anser caerulescens*) in the Northern Arctic region can be either white or dark brown. Eleonora's Falcons (*Falco eleonora*) in the Mediterranean Sea can occur in a slate gray morph or can be orange-colored on the bottom. The females of the butterfly Silver Washed Fritillary (*Argynnis paphia*) can be yellow ocher or green-black. The latter bear the name "forma valesina" and are the spitting image of another species, namely the Cardinal (*Argynnis pandora*). The different intraspecific types have, in part, received additional taxonomic names that are attached to the scientific species or subspecies names, for example, *variatio*, *forma*, and so on. Ernst Mayr also used the term "phenon," meaning a phenotypically varying sequence within the species (Mayr and Ashlock, 1991).

The multitude of terms used in taxonomy, such as variation, aberration, form and phenon, are unnecessary, especially because these terms are used by different authors with varying meanings. It is certainly unclear what is actually meant by *aberratio*. Three terms would be sufficient to designate intraspecific types, thus unequivocally labeling the polymorphic diversity within a gene-flow community. The three terms are race, morph and mutant. Any further designations are superfluous for mutual communication, as well as for the understanding of their evolutionary origin or relevance.

5.4 What are Races or Subspecies?

The term "race" is synonymous to "subspecies." The concept of a race in biology is frequently misunderstood (Sesardic, 2010). To understand the significance of the term race, it is more meaningful to first define what a race is not. Races are not groups of individuals that are phylogenetically closely related, whereas more distantly related groups would be a discrete species. It is a common misperception of species and race to consider them as concepts dependent solely on the phylogenic distance.

If this were true, every species that was evolutionarily old enough could be divided into races. However, there are several species for which races do not exist, although there are populations that have a certain phylogenic distance. This awareness elucidates that there must be something more than the phylogenic distance that allows a population to be defined as a race. These are "race-specific" trait differences that allow a diagnostic separation of the race from the rest of the species. If there are no diagnostic race attributes, then there are no races. Such race attributes are a type of "license plate."

This argument makes clear that, in contrast to a species, a race is typologically defined. It completely depends on a human sorting effort regarding whether some characteristic traits are selected and are considered to declare a subpopulation of a species to be a race. As defined purely typologically, a race is a class (Chapter 3). Races do not exist as such in nature; instead, they are mental constructs that depend on human perception and decision. There are almost unlimited numbers of possibilities for human classification efforts to single out a number of individuals within a species by the similarity of one or a few particular traits. The decision which of these thousands of subgroups within a species is accepted to be a race, depends on a human-made measure of value. Races as distinct groups are not provided by nature but instead only reflect the human principle of order.

Races are subpopulations that differ from one another by specific adaptations to the local environment. Well-known races are human races that are mainly distinguished from each other by skin and hair color and by the structure of the face. Skin and hair color are clear adaptations to the intensity of solar radiation (Rosenberg *et al.*, 2002). Because there are only weak reproductive barriers between individuals of different races, it is a defining characteristic of the race that its individuals merge if they come in contact. Races would disappear if individuals of different races lived in close proximity. Therefore, races can exist only under the necessary condition of geographical distance.

Races can only arise and be maintained if the geographical distance exceeds a certain limit. Otherwise, races would merge. Across short geographical distances, gene flow keeps the members of a population homogeneous, because the genomes of the individuals fuse from time to time by sexual contacts and genetic recombination. Diverging evolutionary pathways are not possible among nearby living individuals because newly arising deviant traits in the individuals at a certain location are continuously reversed by backcrossing with the original organisms (see Section "Why do the individuals of a species resemble each other?" in Chapter 6).

When, however, a certain geographical distance has been exceeded, the gene flow between the distant organisms becomes weaker. A weak gene flow means that the alleles of an individual only rarely reach a distant individual of the species. Under these conditions, gene flow no longer manages to prevent divergence in trait evolution. Therefore, adaptations to specific local environments can establish themselves in geographically distant populations. Such local populations are called races or subspecies.

Races cannot occur at the same location alongside each other because they would merge under these conditions. Thus, races exist always at a geographical distance. The existence of races over a long duration under syntopic conditions is not possible. Neighboring races overlap, but always only in fringe areas. If races move and are then no longer geographically separated, as currently occurs with human races, then the races will merge and disappear in a foreseeable timeframe. The immigration of one race into the residential area of another one does not have an evolutionarily long duration. If a merger between races does not occur, then the resultant scenario relies on the existence of reproductive barriers. In these cases, the populations are not races; they should then be called species.

At the present time, because of mobility and globalization, the human races predominantly are not geographically separated anymore. Therefore, they merge increasingly. In the future, human races will obviously more and more disappear. This scenario is a different form of disappearance of races than the forthcoming extinction of some Indian races in the rain forest of Amazonia. If single human races were to survive in the future, because of sexual barriers that are genetically based or are a consequence of ethnic or religious traditions, then these human races, consequently, would have to be called species.

Between the races, there are smooth transitions in the overlapping regions where the races are in contact geographically. These regions of smooth transitions are called the "clinal" connections between the races. A clinal transition refers to a gradual transition from one geographical race to another, without there being a sharp border. In the transition regions, there exists an almost unlimited merging of the members of different races with each other, because there is not any assortative paring, or there is only a minor amount (Chapter 6). Therefore, in the region of the clinal transition, there live intermediary types between the two neighboring races.

An impressive example of distant populations adapting to local conditions and then differing with regard to their traits is the well-known House Sparrow (*Passer domesticus*). This example also documents how quickly adaptations, and thus the formation of races, can occur. The original notion that the formation of races was a process that lasted millennia had to be thoroughly revised (Johnston and Selander, 1964).

The House Sparrow was originally a Palearctic bird species. In the middle of the nineteenth century, the first specimens were introduced into North America and have then, at the beginning of the twentieth century, spread across the entire northern continent. In the course of only about a hundred generations, considerable local adaptations evolved with respect to coloration, beak and wing size, distinguishing individual races from each other. It would be of special interest to know whether the distant races can actually still cross-breed successfully for a long-ranging number of generations. It is possible that the expected intermediate phenotype of the F1-descendants of a cross-breeding of the distantly living organisms would have lost its ability to successfully compete with pure-bred individuals.

5.5
Are Carrion Crow and Hooded Crow (*Corvus corone* and *C. cornix*) in Eurasia and the Guppy Populations on Trinidad Species or Races?

Often, the question arises of whether two populations are races or species. An example is the Eurasian species pair the Carrion Crow and the Hooded Crow (*Corvus corone* and *C. cornix*) (see Section "The origin of reproductive isolation through reinforcement" in Chapter 6). The breeding areas of the Carrion Crow are geographically separated from those of the Hooded Crow, but both encounter each other in a narrow region of contact, where they successfully hybridize with each other. Geographical separation and clinal transition are criteria for classifying both crows as races. However, the region of contact on the border of the entire geographical range of the Carrion and Hooded Crow is only approximately 50 km wide (Haas and Brodin, 2005). This hybrid zone stretches from Northern Europe across the Elbe region and Austria to Italy. Furthermore, the mating preferences between the two Crows are not arbitrary, but they are clearly assortative (Saino and Villa, 1992). This scenario means that, as a first choice, a Carrion Crow mates with a Carrion Crow, and a Hooded Crow mates with a Hooded Crow. Mixed pairings occur only as a second choice, if the conspecific partner is not found, for example, because, at a slightly later time in the breeding period, most partners are already paired up. The offspring rate of mixed matings is slightly lower than in purebred couples, indicating a postzygotic restriction. Furthermore, the two different Crows have a different choice of habitat, by the Hooded Crow rather than the Carrion Crow preferring stubble fields and corn fields after harvest (Randler, 2008). Therefore, the criteria for classifying both Crows as species exceed the criteria for considering them as races.

At this point, it should be added that there is no scientific gain to waving away the Carrion Crow/Hooded Crow problem with the remark that this case is still an ongoing speciation process that simply has not finished yet. This notion originated in the purely human need for stability, which is not, however, constituted in nature. Everything flows, sometimes more slowly and sometimes faster, and every species is part of a still ongoing speciation process that has not finished yet.

Old historical data speak in favor of considering the current situation of the coexistence of the Carrion and Hooded Crow to be rather stable. It is probable that a similar coexistence of the Carrion and Hooded Crow has already existed for centuries and will continue to exist in the same fashion in the coming centuries, also (Haas and Brodin, 2005; Saino and Villa, 1992). None of the arguments support the view that both crows would become completely separated in the future. In biology, it is more appropriate to look for reasons that allow a partially cross-breeding species pair to continue to remain distinct in spite of mutual hybridization, than evading the problem and speaking of a temporary situation.

Another borderline case between a race or species is provided by Guppies (*Poecilia reticulate*) on Trinidad (Magurran, 1999). The Guppies are another example of the fact that a large degree of diversification can successfully coexist with ongoing gene flow of long duration between different populations. Diversity among populations should not be considered as necessarily leading to a coming speciation. Speciation is not the inevitable consequence of a large degree of diversification in a population.

The Guppies that are widespread over the inshore waters of Trinidad split up into a multitude of small populations that have different traits. These trait differences surpass the diversity of well-established species. Molecular analyses have revealed that the different individual populations have existed for half a million years, which is more than a million generations. In spite of this extent, the different populations are reproductively connected to each other in broad transitions. There is neither a homogenizing merger of the population differences nor is there an erection of reproductive barriers such as in the case of another family of fish, the African Cichlides, which are subject to rapid speciation (Chapter 6). It is an unresolved mystery why certain groups of animals form many new species in evolutionarily short times whereas others, sometimes closely related groups of animals, do not form any new species over long periods of time (Seehausen *et al.*, 2008; Verheyen *et al.*, 2003).

Why is there no speciation in the case of the Guppies? Possibly, the rigorous pursuits, in which the Guppy males "conquer" the females for reproduction, play a role. This behavior, to a great extent, suspends the usual biological mechanism of "female choice" (Magurran, 1999). Normally, the selective choice of males by females is the main reason why speciation occurs. In most higher animal species, males that deviate from the norm are not allowed to reproduce by the females, which can lead to the formation of species barriers that then remain in existence. In most species, this choice and decision behavior is initiated by the female. It is called "female choice" (Chapter 6) and is one of the decisive forces that leads to species formation. The males act against this force. In the case of the Guppies, the males breach these barriers and, thus, maintain the gene flow even between rather different populations. The main information that can be learned from the Guppy example is that population diversity within a species can be evolutionary stable, despite ongoing gene flow.

5.6
What are Morphs?

Another example of the fact that diversity within a species can benefit the species itself and should definitively not be thought of as leading to a speciation is provided by the morphs. Similar to races, morphs are examples of intraspecific polytypes. However, there is a clear difference between morphs and races. Morphs do not merge to become intermediate types if they are crossed; instead, their offspring again split into distinct morphs. Races, in contrast, produce intermediate types if they are crossed.

The most commonly known example of morphs is the sexual dimorphism, the simultaneous existence of females and males. In several species, the two sexes are phenotypically (not genotypically) very different. Already at the morphological level (hence the name dimorphism), enormous differences between males and females within a single species become apparent. With many animal species, it is easier to distinguish the males from the females than to differentiate between different species (Color Plates 5 and 6). An example of this scenario is the ducks, whose females can often be assigned to species only by specialists, whereas the distinction between males and females of the same species is not difficult for most of the year.

If only two morphs exist, then there is a "dimorphism;" if more than two exist, then there is a "polymorphism." Morphs are different phenotypes of the same species that often live at the same location and that reproduce in an unlimited fashion with each other. Morphs should not be confused with races. It is an important difference between races and morphs that races maintain their phenotypic integrity because they are, to a great extent, isolated by distance, even though there are smooth transitions in geographic regions of contact. In great contrast, morphs, in most cases, occur alongside each other syntopically and are not limited in any way from mating with each other. This scenario raises the question of how morphs maintain their phenotypic integrity, because, in contrast to races, there is no mating barrier by geographic distance, and, in contrast to species, there is no mating barrier by intrinsic traits that prevent sexual merging.

Why are there no intermediary phenotypes between morphs in spite of a strong gene flow? Why is the result of a crossing between a male and a female not an intermediate sex? Morphs owe their polytypy to specific mechanisms that prevent phenotypic commixture. In most cases, these mechanisms are genetically based. Specific genetic mechanisms prevent the formation of intermediary phenotypes, with the best-known being the example of sexual dimorphism. However, in some cases of di- or polymorphism, intermediaries arise without being prevented by genetic control. In these cases, selection dominates. Although organisms that are intermediates between two distinct morphs are born, they have only limited chances of growing up. Only a few distinct morphs grow up. The intermediates constitute no optimal adaptations to any type of existing environmental conditions. Therefore, they are weeded out by selection more or less shortly after their origin in every generation and, thus, hardly make an appearance. One of the best examples of this case is the Garter Snake *Thamnophis ordinoides* (see below).

One must distinguish morphs that are solely alternating developmental stages of an organism from those morphs that have the same developmental age and coexist at the same time of the year alongside each other, for example males and females. Only the second type of morph is a true morph. An example of the first type of morphs is larvae, in contrast to imagos. Larvae sometimes live in a completely different environment from the adults and often are adapted to food sources that differ from food of the adults. Specifically in holometabolic insects, caterpillars and pupae do not resemble mature butterflies of their own species in most of the phenotypic traits.

Some organisms, such as Trematodes (parasitic flatworms), several Coelenterates (polyps and jellies), and plants, have alternating generations in their life cycle. In some cases, diploid stages alternate with haploid stages. In other cases, stages that reproduce vegetatively alternate with stages that reproduce bisexually. Some butterfly species are characterized by bivoltinism or even multivoltinism (see Chapter 6 and Color Plate 1). Multivoltinism means that two or even more reproducing generations succeed each other in the cycle of a year; the first generation hatches in the spring, and the second or third generation of butterflies hatches in early or late summer. These two or more generations may differ in wing color and wing pattern. The example most commonly known in Europe is the Map Butterfly (*Araschnia levana*), whose wing color in spring is mainly orange, whereas it is black in summer.

All of these examples of polymorphic stages in the course of development or in the sequence of generations, however, are not true polymorphisms, because they are different developmental stages and are only expressed phenotypically and do not reflect differences in the underlying genotypes. True polymorphisms are morphs whose polytypy is founded on a genetic polymorphism. The differences in phenotypes are founded on different genomes.

This scenario is the case with many forms of sexual dimorphism. Aside from sexual dimorphism, there is ecological polymorphism. Different morphs of the same species are adapted to different ecological niches or feed on different plants or prey. Then, there is also the mimicry polymorphism. Different morphs of the same species mimic entirely different animal species, to be protected in different forms against predators (see Color Plates 2 and 3). Nevertheless, they all belong to a single species. In all of these cases of true polymorphisms, the genetic basis for the existence of distinct morphs is a stable allelic polymorphism (see below).

For taxonomy, these examples of polytypy make clear that you cannot infer the presence of a different species from different appearances. This relationship shows that a purely phenetic species concept (Chapters 2 and 4), which treats species differences as nothing more than trait differences (Sneath and Sokal, 1973), cannot be maintained consistently. In the case of sexual dimorphism, there is no longer the possibility of determining a species by what looks different. According to the pure criteria of phenetics, two dimorphic sexes would be declared to be two different species, simply because they differ in several traits (see Color Plates 5 and 6). To avoid this inaccuracy, a second species concept must be consulted in addition to the phenetic species concept, namely the concept of the gene-flow community. Otherwise, a system with biologically little meaning would be constructed.

Sexual dimorphism, however, does not involve only morphology. Sex-specific traits also affect metabolism and brain function, behavior, food utilization, reactions to specific environmental stimuli and the patterns of transcribed genes in specific tissues (the transcriptome), as well as the expressed proteins (the proteome). All of these traits can differ between the two sexes. With closely related species, all of the differences between the sexes can significantly exceed the differences between two different species (Billeter *et al.*, 2006).

That a human at first spontaneously thinks of different species when he perceives the different appearance of males and females is proven better by no other than Linnaeus himself, who at first took the two sexes of the Mallard (*Anas platyrhynchos*) to be different species and gave them different names (Mayr, 2000). Linnaeus also described the immature Goshawk (*Accipiter gentiles*) as a distinct species because of its vertically striped belly plumage, as distinguished from the mature Goshawk, which has horizontally striped belly plumage (Mayr, 2000). Thus, nature has not shaped species for humans to be able to identify them.

5.7
What are Mutants (in a Taxonomic Sense)?

In addition to races and morphs, there is also an additional intraspecific "morphotype," which, however, in contrast to the race and morph, does not have any taxonomic importance. It must be considered to be a genetic accident, has selective disadvantages, and is, thus, usually soon "weeded out" again. What is referred to here are the occasionally occurring aberrant mutants, such as, for example, the albinos, a partly or completely colorless phenotype, resulting from an inherited dysfunction in the biosynthesis of the melanins. Albinos as genetic accidents should not be confused with the white morphs occurring in the case of certain animals, for example, several heron species (see above). These types have other genetic causes, and obviously play an evolutionary role and should not, therefore, be considered to be "accidents."

The term mutant is defined very broadly in genetics and usually designates any form of genetic alteration in the genome, from a point mutation up to a chromosome or genome mutation. For taxonomic classification, by far not all mutants are important. Only those mutants that produce diagnostically clearly recognizable phenotypes are relevant. Only such mutants are meant in this context.

Mutant that distinctly alter the phenotypic appearance of an organism have repeatedly confused taxonomists. Mutants can feign distinct species (Color Plate 7). The spectrum of such mutants ranges from diverging colors and color patterns up to deformed shapes, such as the so-called "mongoloid," a human carrier of "Down syndrome," which results from the presence of three copies of chromosome 21.

Melanism occurs frequently in very different animal species. This phenomenon refers to abnormal types with dark to black body color, as can, for example, be observed in the case of various mammals, but also other animals. Melanism is based on an excessive deposition of melanins. Especially famous are the melanistic mutants of several species of Felidae (cat-like animals) and especially the case of the Black Panther, a mutant of the Leopard (*Panthera pardus*). The genetic cause for melanism in the case of the Leopard is a single gene that is inherited recessively (O'Brien and Johnson, 2007). Interestingly, Black Panthers occur in different geographical regions in substantially varying frequencies. Because melanism in the Panther is a recessive mutant, a positive selective advantage can be assumed. If Black Panthers are indeed promoted by selection, then the Leopard would have to be considered to be a polymorphic species, and the melanistic Panther should then consistently be designated as a morph, not as an insignificant mutant.

5.8 Allelic Diversity

Intraspecific polymorphisms appear in different forms. They start with the alleles of any single gene being able to mutate and, thus, become present in different variations in a population's genomes. The organisms of a population are, thus, fairly different. This phenomenon is called multiple allelism. In the case of humans, two arbitrarily chosen haploid genes on average show a difference within every 1250 bases (Venter *et al.*, 2001). This variability means that most genes are present heterozygotically and that, therefore, the individual organisms of a population all differ from each other. However, in comparison to his closest relative, the chimpanzee, the human is genetically fairly homogeneous. The individuals of the chimpanzee differ even more from each other. The reason is that *Homo sapiens* went through a narrow population-genetic "bottleneck" when emigrating from Africa 40 000 years ago and, thus, lost many alleles due to genetic drift. With *Drosophila melanogaster* and many other animals, the mean value of the heterozygotic difference of nucleotides is even significantly higher than for humans (Aquadro, Bauer DuMont, and Reed, 2001). With *Drosophila melanogaster*, even small genes of only 1000 base pairs that are homologous to each other on average differ at four positions per gene, as long as the genes are independent from each other (Powell, 1997).

The inference of the phenotype from the genotype is admittedly difficult. Most phenotypic differences between closely related species are not the expression of differences among protein-coding structural genes (Chapter 4). The development of the body plan is, instead, controlled by tissue-specific transcription factors. If a transcription factor is not expressed at the correct time or a tissue-specific enhancer is mutated, then differences in body plan, physiology and behavior may occur (Prud'homme *et al.*, 2006). This scenario usually is not caused by the fact that individual structural genes, on which the traits are based, are altered, but by the fact that the individual structural genes are regulated differently, even though they remain unchanged themselves.

Every organism is a singularity in its complexity. Thousands of traits differ between single organisms within the same population. What, then, is an intraspecific difference in comparison to an interspecific difference? When do trait differences equate to species differences, and when do they indicate only intraspecific variability (Chapter 4)?

The extent of intraspecific variability depends on the population size, because inbreeding quickly reduces the allelic diversity. All species that have passed through a genetic bottleneck, have less allelic diversity, because the number of individuals was severely reduced by the bottleneck. Their genomes are, to a great extent, homozygous. Examples of genetic bottlenecks are several cat-like animals (*Felidae*) that were severely persecuted and whose territories were carved up. This scenario is the case for the Cougar (*Puma concolor*) in North America; cougars now occur in a few isolated geographical ranges in the US, in each case with only a small number of individuals, for example, in Florida (Roelke, Martenson, and O'Brien, 1993). The Cheetah (*Acinonyx jubatus*) had at one time spread from Africa far throughout the Middle East to India. It already, apparently, went through a bottleneck centuries ago, so that all of the specimens living today in Africa descend from a very small stem group (Harvey and Read 1988).

5.9
How Long is the Lifetime of Allelic Polymorphisms?

Already Students in introductory genetics classes learn that mutant alleles are rare exceptions to the standard wild-type allele. This is basically correct given that mutations generally result in decreased fitness; most mutants are therefore quickly weeded out by selection. In rare cases where a mutant allele increases fitness, this allele generally replaces the wild-type allele in the entire population within only a few generations, that is, the mutant allele becomes "fixed" in the population. Consequently, given that selection promotes the survival of only the fittest variant, any multiplicity of allelic variants is generally short-lived, at least in small populations.

However, in addition to allelic variations that are controlled by selection, some are selectively neutral. Neutral alleles result in neither phenotypic advantages nor disadvantages. These alleles therefore are not eliminated or supported by selection and, because new mutations keep occurring, should exhibit a continuously increasing frequency in a population. However, this is not observed to be the case. Neutral allelic variations also become lost with time. This is due to genetic drift, which over certain timeframes leads to the disappearance of many allelic variants. This process is particularly rapid in small populations. If one assumes that no new mutations emerge over a long evolutionary timeframe, allelic diversity would then gradually disappear via genetic drift, and the genomes of the entire population would become homozygous.

This process can be explained in a simple analogy (Cavalli-Sforza and Cavalli-Sforza, 1994). Assume that many individuals in a single population all bear different names. Assume that the offspring of a marriage receives the name of only the father. As a consequence, the name of everyone who does not produce children or who is a woman becomes extinct. Accordingly, the diversity of names decreases from one generation to the next. This process has nothing to do with the names being cumbersome or even harmful, they disappear purely by chance. Eventually, the entire population may only bear a single name. This process is precisely what is meant by genetic drift in population genetics. A diversity of names can be maintained in only two ways: immigration from alien populations or the reassignment of names. The latter would be analogous to mutations in the context of genetics.

In contrast to natural selection, where only positive alleles have a chance at survival, in the context of genetic drift, the extinction or fixation of an allele in a population is a matter of chance. It therefore follows that in every population, allelic diversity is short-lived and only exists because of the constant addition of new mutations.

5.10
Stable Polymorphisms – The Selective Advantage is Diversity

The principle that allelic diversity is evolutionarily short-lived is, however, upset by a remarkable exception: stable polymorphisms. Stable polymorphisms refer to the long-term persistence of different allelic variants within a population, without the elimination of single variants by selection or drift.

Selection normally favors only one of several allelic variants because only one variant is optimally adapted to the specific task of the given gene. The other allelic variants are most likely less able to perform this task and therefore should have a short lifetime.

Stable polymorphisms, however, are a remarkable exception. How can several variants of a single gene survive for a long period of time in a population? Individual allelic variants cannot have equal fitnesses. In the case of stable polymorphisms, the selective advantage for allelic survival appears not to be the quality of the individual allele but rather the coexistence of several parallel allelic variations in the population; that is, selection is "interested" in multiple allelic forms coexisting within the population. Such a population is composed of individuals that carry different allelic variants of a gene. Therefore, the population always maintains a number of organisms that are adaptively prepared for potential environmental changes. In this way, the population can react more flexibly to varying environmental changes. Such allelic variations can therefore be said to be "kept in reserve" by the population to be recruited in a vital moment. Such maintenance of variation is referred to as preadaptation.

The selective advantages of allelic diversity in a population, as well as preadaptation, cannot be easily understood from a Darwinian perspective. Ultimately, however, individual organisms benefit from the populations' advantage. This concept amounts to group selection, which is rejected by many evolutionary biologists. The relationship between individual and group fitness is a controversial matter in the field of biophilosophy. Multiple allelisms and polymorphisms are a matter of group fitness. Theorists, however, disagree whether natural selection acts primarily on genes, individual organisms or whole populations (Okasha, 2003; Okasha, 2006; Okasha, 2009).

It is argued that only entities that replicate and reproduce, that is, leaving offspring, can be units of selection and that a necessary condition of reproduction is that the offspring outlives its parent (Dawkins, 1976). Does a population reproduce, or do only the single organisms of a population reproduce? The group selection controversy concerns whether natural selection can operate at the group level rather than at the level of individual organisms. Most contemporary evolutionary biologists are highly skeptical of the hypothesis of group selection, which they regard as biologically implausible (Okasha, 2001).

How can an allele survive the long term in a population if it is not advantageous, or even disadvantageous? Such an allele would certainly be selected against relatively quickly and thus disappear in the long term. As far as allelic polymorphisms are understood today, many are not immediately advantageous for the individual. Instead, the advantage for survival lies in the population itself surviving in the case of an environmental change. In the case of such an environmental change, the survival of a population can depend on the few surviving individuals that carry alleles that were previously disadvantageous (see the example of Darwin's Finches below). However, how do organisms (and their alleles) that are disadvantaged survive in an unchanging environment?

This difficulty is not simple to resolve. Which genetic mechanisms allow for the survival of occasionally disadvantageous alleles? One mechanism is immediately obvious: recessivity. Many alleles, even disadvantageous ones, can survive for a long period of time if they are expressed in the heterozygous state. This allelic survival is possible because the homologous dominant allele determines the phenotype and is therefore the only allele that is exposed to selection. Such recessive alleles are under a "invisibility cloak." Certain authors even believe that the biological meaning of diploidy is founded in this concept.

A second genetic mechanism that may preserve disadvantageous alleles over the long term is the fixed coupling of these alleles with neighboring genes. These groups of genes would be linked on the chromosome and only rarely, if ever, separated by recombination. If the neighboring genes have a selective advantage, they could carry disadvantageous coupled genes with them from one generation to the next, provided that the disadvantage of the linked allele is not too great.

The following simple example further illustrates this concept. Sexual dimorphism can be thought of as a stable polymorphism. There are several examples of a particular sex exhibiting selective disadvantages. For example, there is a clear selective disadvantage for the female sex in some human cultures. Despite this, females are not displaced through selection because the selective advantage of sexual dimorphism is not the advantage of being a male or a female. Rather, the advantage comes from there being two sexes in the population. The sex that is disadvantaged by inherited properties or unfavorable ethical or social constraints does not have a selective disadvantage; its survival is due to the advantage of the two sexes coexisting, as disadvantageous as it may be to be of a given sex.

Stable polymorphisms are not restricted to sexual dimorphism. These polymorphisms occur in various forms and are the primary foundation of intraspecific trait variabilities (Aguilar *et al.*, 2004). Known examples of stable polymorphisms include the major histocompatibility gene complex (MHC) and the different blood groups. In the case of the MHC, allelic diversity is evolutionarily older than the separation of humans and chimpanzees. With respect to particular MHC alleles, humans and chimpanzees are more similar than particular humans are among themselves (Figueroa, Günther, and Klein, 1988). Careful thought should be given to the fact that certain alleles carried by two members of different species can be more highly related than alleles carried by two members of the same species.

While polymorphisms within the MHC alleles predate the speciation of higher primates, the allelic polymorphism of blood groups may have convergently evolved in primates. The A, B and O groups exist in both humans and rhesus monkeys; in contrast, only the A and O groups are present among chimpanzees. In contrast with earlier opinions, this polymorphism does not predate the divergence of the different primate species. With all probability, blood groups do not represent an example of an ancient ancestral allele but a polymorphism that has apparently repeatedly evolved following the divergence of the different primate species (Kermarrec *et al.*, 1999). It follows that the polymorphisms leading to the different blood groups must have a strong selective advantage in a given population (i.e., the advantage does not lie with a single allele). This can be concluded because these polymorphisms did not disappear in favor of one of the three alleles A, B or O, but repeatedly originated in different species.

Stable polymorphisms impede trait-oriented taxonomic classification and are generally an obstacle in taxonomy. Many species are polytypic. The adherents of the phenetic species concept face the significant problem of dealing with the existence of definable groups within a species. It must be concluded that stable polymorphisms are of importance to not only taxonomists but also in everyday life. Biological diversity is not only expressed by species diversity but also by intraspecific diversity (Lockwood and Pimm, 1994). This fact has consequences for the objective of species protection projects (Chapter 2). If "species conservation" is defined literally, that is, the protection and preservation of species, then biodiversity is not truly being protected (Moritz, 1994; Crandall *et al.*, 2000; Smith, Schneider, and Holder, 2001; Allendorf and Luikart, 2006).

5.11
Are Differences between Species Due only to Differences in Allelic Frequency Distribution, Such that there are no Truly Species-Specific Traits?

An important parameter within a population is the quantitative distribution of different alleles within different populations. For example, assuming that an allele is present in a population in four different variants, then any one of the four alleles should represent 25% of the variants in each population. In natural populations of animals and plants, however, this is not the case, and the allelic distribution diverges from the expected uniform distribution. Even if all of the variants of a particular allele are present in all of the populations of a species, their frequency distribution distinguishes the populations from one other. In addition, the allelic distribution changes within the population over time.

For example, the frequency distribution of the alleles that define the ABO blood groups differs significantly among human populations (Mourant, Kopec, and Domaniewska-Sobczak, 1976). Nearly 99% of the native population of Latin America are type O; in contrast, within certain Central African populations, only 30% are type

O. However, even between nearby populations, the frequencies of the blood group alleles can differ considerably. For example, in Germany, Bavarians exhibit a significantly different distribution of blood groups than Westphalians (Mourant, Kopec, and Domaniewska-Sobczak, 1976).

Because most alleles are widespread, genetic differences among human populations derive primarily from gradations in allele frequencies rather than from distinctive "diagnostic" genotypes. Indeed, many populations differ from each other only in the accumulation of allele-frequency differences across many loci (Rosenberg *et al.*, 2002).

The average proportion of genetic differences between individuals from different human races only slightly exceeds the differences observed between unrelated individuals from a single race. That is, the within-race component of genetic variation accounts for the majority of human genetic diversity. This overall similarity within human populations is also evident in the geographically widespread nature of most alleles. Most alleles are present in all human races: African, European, Middle Eastern, Central/South Asian, East Asian, Oceanian and American. Fewer than 8% of all alleles are exclusive to a single region (Rosenberg *et al.*, 2002).

The Hardy-Weinberg Law states that the allelic frequency distribution within a population is not altered by sexual processes, that is, neither by meiotic recombination nor by a genome merger during fertilization. The Hardy-Weinberg Law, however, presupposes an externally closed and in every respect constant population. This is never the case in nature; the allelic frequency distributions of natural populations exhibit spatiotemporal changes. Such changes are due to the following five factors:

1) Mutations transform one allele into another.
2) Individuals invade from alien populations, thus changing the allelic frequencies.
3) Assortative mating occurs, that is, the gametes in a population do not randomly meet to form zygotes, but particular sexual partners are preferably selected. The genomes of these individuals are reproduced more frequently than those of other organisms.
4) There is varying vitality and reproductive success among the individual organisms, such that certain alleles have a greater chance of being inherited.
5) Genetic drift occurs, during which certain alleles randomly become rarer or more common.

Differences in allelic frequency between two populations are a serious issue for taxonomists. Let us explain this problem using the example of the AB0 blood groups. First, assume that 50% of the individuals in one population are type 0, and that this percentage is 20% in another population. No one who classifies species according to their trait differences would designate these two populations as different species. However, what if this difference in percentages becomes more extreme? Now assume that 99% of the individuals in the first population are type 0 and that this percentage is 1% in the second population. If a taxonomist collected fewer than 100 specimens, he/she would not recognize that the blood group 0 also exists in the individuals of the second population despite the fact that this trait exists in 1% of these individuals. The taxonomist may argue that the trait "blood group 0" is a diagnostic trait that characterizes the first population only and, therefore, may be a qualitative difference.

Several geographically separated populations may differ only in allelic frequency distribution, not in the absolute presence or absence of a particular trait. The question arises of how taxonomists should handle this issue if the same situation applies to species. Can two populations be accepted as two different species if they carry precisely the same genes but differ only in the frequency distribution of their allelic variants? Can certain species differences (similar to population differences) simply be differences in allelic frequency distribution, without any trait being specific to one species? This would be an unsatisfying situation.

The issue of different allelic frequencies is most striking if these differences are large. Should a new species be named if allele 1 is carried by 99% of the individuals in one population but by 1% of the individuals in another? What if allele 1 is carried by 50% of the individuals in the one population and by 40% of the individuals in the other population? There is no qualitative difference between these two sets of populations.

Allelic frequency differences imply that no single trait is 100% reliable in defining two species. This concept can be illustrated by the following model with respect to the groups *a* and *b*, which are characterized by the traits *A* and *B*:

Six individuals of group *a* have the properties: *A A A A A B*
Six individuals of group *b* have the properties: *B B B B B A*

No single trait (*A* or *B*) constitutes a qualitative difference between the groups *a* and *b*. The traits *A* and *B* therefore cannot be used as reliable species-specific identification traits between the two different groups, although there is undoubtedly a significant difference between the two species with respect to these traits.

It is a difficult question whether differences in allelic frequencies can be accepted as species differences or whether species differences must always be based on trait differences that have 100% penetrance. In most cases of trait-oriented taxonomic classifications, the number of observed samples does not suffice to distinguish between these alternatives. If it is accepted that differences in the trait's frequency distribution indicate species differences, then the above examples demonstrate that a single trait cannot be used to make the distinction.

5.12
Partially Migratory Birds – an Example of Genetic Polymorphisms

A remarkable example of population differences that are based on allelic frequency distributions is the difference between migratory and sedentary birds (Berthold and Querner, 1981). This example, however, is fraught with several unsolved problems. Ornithologists note that the migratory behavior of birds differs among populations of a bird species, and many species consist of two very different geographical populations. The migratory population leaves their breeding habitat during the fall, returning to the breeding habitat again the following spring. In other geographical regions, however, sedentary animals of the same species remain in their breeding habitat during the winter. At least in the example of long-distance migratory birds, this coexistence of migratory and sedentary birds is not due to voluntary decisions of the individual birds but is genetically programmed (Berthold and Querner, 1981; Sutherland, 1988). The two populations of the same bird species are genotypically different.

The migratory disposition of most birds, that is, their motive for leaving their breeding habitat, is not triggered by seasonal climatic changes, at least in the case of long-distance migratory birds. Only for so-called "cold fugitive" species ("Kälteflüchtlinge") are decreasing temperatures or food shortage the immediate triggers for the migration. Among these species are certain European birds, which generally migrate only short or medium distances, for example, many ducks, geese and swans (*Anatidae*) and some herons (*Ardeidae*). Several individuals of these species remain in their breeding habitat in the fall or pause in stopover regions on their way to the southwest of the breeding habitat. These latter individuals remain in these stopover regions until frosts and food shortages cause them to migrate further southwest.

In the case of most long-distance migratory birds, however, it is not simply food shortages or the encroaching cold that causes the animals to leave their breeding habitat for southern or western climes. These migratory birds leave their breeding habitats during high or late summer, when the temperatures and food supplies are still optimal. Consider the European Swift (*Apus apus*), which leaves central Europe in mid July, when airborne insects are still available in sufficient amounts. These birds do not migrate because certain living conditions have worsened or because they are starving or cold. Instead, these migrations are controlled by the decreasing length of the day. This migration behavior has a genetic basis. The birds become restless due to an internal clock, not because of external compulsions.

The genetic disposition for migratory or sedentary behavior likely requires the cooperation of several genes. Nearly nothing is known regarding these genes (Helm, 2009). However, in most cases, the genetic differences between migratory and sedentary birds are likely not differences in structural genes but rather (1) differences within regulatory elements (enhancers) of structural genes and (2) tissue-specific activities of certain transcription factors (Chapter 4).

Migratory and sedentary birds are distinguished by a multitude of traits. A migratory bird requires additional instincts. In migratory birds, genes are active that control the instinctive migratory behavior twice a year at the proper time; these genes are inactive in sedentary bird populations. Migratory birds also require a different metabolism and anatomy.

In contrast to sedentary birds, migratory birds become restless in the late summer and fly continuously in a particular geographic orientation. The direction of migration and the migration distance are genetically determined. This is known because many songbirds migrate at night without parental guidance. These birds must locate the migration paths themselves, without the help of other individuals. Small songbirds are short-lived and can hardly benefit from the experience of the previous year. Instead, the bird must know instinctively whether it must migrate in the southwest or southeast direction. The bird must also know when it has arrived in the wintering grounds. A migratory bird must be genetically programmed to know how many days or weeks to maintain its migratory restlessness before ceasing to migrate. Otherwise, the bird may cease its migration too early or too late and would stop in the wrong winter habitat.

All of these behavioral properties of migratory birds have a genetic foundation. Sedentary birds do not require the instincts needed to migrate; the genes responsible for migratory behavior are inactive in the individuals of the sedentary population of the same species. Furthermore, migratory birds require a different metabolism than sedentary birds. Such birds must build up enormous subcutaneous fat deposits as fuel as well as water deposits before the fall migration. These different metabolic requirements necessitate a multitude of enzymatic activities, which are all repressed or differently controlled in sedentary birds. Before beginning their migration, therefore, migratory birds may increase their weight by a factor of two in the late summer. During these weeks, migratory birds may weigh twice as much as members of the sedentary birds of the same species (Berthold and Querner, 1981). In addition, wing length and wing pointedness are greater in migratory birds than in sedentary birds (Rensch, 1938; Baldwin *et al.*, 2010). Blackcaps (*Sylvia atricapilla*) on the Cape Verde Islands are sedentary birds and have much shorter wings than do blackcaps on the European mainland. The short wings of the Cape Verde Islands population would not allow them to fly across the ocean (Pulido *et al.*, 2001).

The enormous number of different traits between the individuals of migratory and sedentary population of the same bird species raises the taxonomical question of whether migratory and sedentary birds are different races or morphs.

According to the position of Berthold, there are no bird populations that are either completely migratory or completely sedentary; the difference in traits between the migratory and sedentary birds is simply a matter of allelic frequency distribution. Therefore, a bird population that lives in the north and normally leaves its breeding habitat in the fall must contain a few individuals that are genetically determined to be sedentary. In cold winters, these individuals die off.

However, a few alleles survive in the entire population, presumably hidden under the cloak of recessivity. Recessive alleles are not phenotypically expressed and, therefore, can survive without being eradicated by natural selection. The analogous situation can be supposed for sedentary bird populations. The members of this population remain in their breeding areas over the entire year. They do not migrate because they find enough food even during the winter. In this population, however, a few individuals may possess the allelic constitution that encourages them to migrate to distant winter habitats. This behavior in a normally sedentary population may be detrimental given that when these birds arrive in spring, their potential breeding territories may already be occupied. However, migratory alleles will survive in a sedentary bird population.

Berthold's position is also supported by the fact that, if climate changes demand it, a population of migratory birds can turn into a population of sedentary birds, or vice versa, over only a few generations. The allelic polymorphisms in alleles that control migratory and sedentary behavior provide the individual populations with a remarkable flexibility over a longer evolutionary timeframe. A change of the local climate can, within a small number of generations, transform a migratory bird population into a sedentary bird population. The individuals that exhibit the necessary preadaptive traits replace other individuals given their sudden colossal selective advantage (Pulido *et al.*, 2001). Because the genes that control migration theoretically survive in some individuals in the sedentary population, the entire population of sedentary birds preserves the opportunity to become migratory in the event of a climate shift. The analogous situation is true in populations of migratory birds; if the environmental conditions change, then a population of migratory birds can, via selection, become sedentary after a small number of generations.

A historic example of such a transition is the Eurasian Blackbird (*Turdus merula*), a sedentary bird in Central Europe. However, two hundred years ago in Germany, in Goethe's times, this blackbird was migratory and was not observed in the winter. During this period, the winters were much colder, particularly in the cities.

The question arises of whether migratory birds and sedentary birds are races or morphs. Because Eurasian migratory populations generally breed in the north, and sedentary populations generally breed in the south, the two populations live in specific geographical regions. This coexistence of two different populations as a local adaptation to different geographic environmental conditions suggests, at first glance, that these populations are geographic races (see above).

However, there are geographical regions in which migratory and sedentary birds of a species live side by side. In such clinal regions of migratory and sedentary birds, the two populations interbreed. In these situations, it can be directly tested whether the two populations are races or morphs. If the populations are races, the mixed population should consist of intermediary phenotypes, indicating that their genes recombine. That is, a portion of the "migratory alleles" should recombine with a portion of the "sedentary alleles." The result of this, however, would be expected to be lethal given that a bird carrying half of the migratory alleles and half of the sedentary alleles would not survive. However, contrary to expectations, where migratory and sedentary birds overlap in the clinal transitory region, the mixed populations constitute a viable reproductive coherence. These populations do not in fact produce predominantly lethal hybrids but rather two very distinct phenotypes: viable migratory individuals and viable sedentary individuals.

Consequently, the birds that live in the regions where migratory and sedentary populations of birds overlap are referred to as partially migratory birds. The migratory portion of this population leaves the breeding habitat in the fall and returns in the spring of the following year. The sedentary portion of the same population in the same region does not migrate.

How can this be? The numerous hereditary dispositions of migratory and sedentary birds are certainly mutually incompatible. It should be expected that migratory and sedentary birds cannot successfully crossbred with each other if their genes recombine. If the two populations breed, it should be expected that the intermediate hybrid is partially equipped with "migratory allelic constitutions" (i.e., alleles that control migratory restlessness, orientation and migration distance, fat metabolism, wing dimensions, etc.) and partially with alleles that confer sedentary traits.

That the breeding of these two different phenotypes does not result in intermediate phenotypes but rather in two distinct phenotypes is precisely what characterizes the concept of a morph (see above). Morphs are distinct complex types that interbreed without producing intermediates, instead producing distinct phenotypes in an either-or fashion.

Therefore, migratory and sedentary birds must be considered morphs, comparable with the formation of males and females in the example of sexual dimorphism, where no intermediate types appear. If migratory and sedentary birds are considered races, the traits for migratory and sedentary behavior should merge gradually. The crossing between migratory and sedentary birds in the clinal transitory region, however, generates predominantly either migratory or sedentary birds.

Most investigations of migration polymorphisms in birds have been performed on Blackcaps (*Sylvia atricapilla*) (Berthold and Querner, 1981) and Stonechats (*Saxicola torquata*) (Helm, Fiedler, and Callion, 2006). Blackcaps have a large breeding area that ranges from the Cape Verde Islands (on the same latitude as Senegal), the Canary Islands and throughout nearly all of Europe. Populations are also observed in Iran and in the Ob region of the West Siberian Plain beyond the Ural Mountains (Chapter 6). On the Cape Verde Islands, Blackcaps are sedentary birds. However, in the entirety of southern Europe, both migratory and sedentary birds can be found. In northern Europe, Eastern Europe and Asia, Blackcaps are migratory. in Europe, Stonechats are fully migratory in Germany and in the Benelux states, but British stonechats are partially migratory; nearly half of the British Stonechats migrate, whereas the other half is sedentary (Helm, Fiedler, and Callion, 2006).

How are migratory and sedentary birds able to interbreed with long-lasting success in a partially migratory population? How is it possible that migratory birds can successfully cross with sedentary birds? One line of thought is that there is a genetic mechanism that protects entire complexes of genes from recombination, maintaining their linkage and preventing crossing-over. In this situation, recombinants of the linkage groups would not occur in the partially migratory population. That is, the decision would be either-or in this population, similar to many examples of the genetic determination of sex. Another line of thought would hold that all "migratory genes" are controlled by a master gene that is either switched on or off, thus regulating an entire set of subordinate genes.

Recent investigations, however, indicate that the assumption of a distinct segregation of bird individuals into residents and migrants in the clinal overlapping region of migratory and sedentary populations has been overestimated (Helm, Fiedler, and Callion, 2006). There intermediary hybrids exist that travel over a continuous range of distances into varying wintering grounds. Unfortunately, experimental data regarding crossings between migratory and sedentary birds are limited, requiring further substantiation of the classification of migratory and sedentary birds as morphs.

5.13 Intraspecies Morphs in the Burnet Moth *Zygaena ephialtes*

Many species of the butterfly genus *Zygaena* (Burnet Moth) can be distinguished from each other phenotypically only with difficulty. These species are therefore referred to as cryptic species. An example of these are the three species of the *Zygaena-transalpina* group: *Z. transalpina*, *Z. hippocrepidis* and *Z. angelicae*. Only specialists are able to distinguish these three species from each other, and an element of uncertainty remains even then.

Alternatively, only in the genus *Zygaena* can an impressive example of intraspecific polymorphisms be observed, for *Z. ephialtes*. This species consists of a number of morphs that can be distinguished from each other much more easily than the majority of the other *Zygaena* species can be differentiated at the species level (Color Plate 4). Anyone who intends to identify *Zygaena* species can only dream to be able to identify individual species as easily one can distinguish the individual morphs *Z. ephialtes*.

Zygaena ephialtes consists of four different morphs. Therefore, *Z. ephialtes* also bears the trivial name variable Burnet Moth. The drastic phenotypic differences between the morphs are based on only two genes, for which there are each two allelic variants with two different phenotypes. One gene regulates the phenotype of the hind wings, which are either colored or black. The other gene regulates the color of the moth, which can be either red or yellow (Sbordoni *et al.*, 1997).

The alleles of the two genes act in a dominant-recessive fashion. Colored hind wings are dominant over black wings. Red color is dominant over yellow. Both genes can be freely combined, because they are located on different chromosomes. This very simple genetic constitution explains why four different morphs arise.

Zygaena ephialtes is one of the best examples of very different phenotypes that simulate different species. These differences, however, are based on only two genes, and the entire phenotype set is a textbook example of the third Mendelian rule. The genetic crossbreeding of homozygous parents corresponds to two-factor crossing, with the well-known result that four phenotypes occur in the F2 generation in the 9:3:3:1 ratio. These four phenotypes are the four morphs of *Z. ephialtes*: a red morph with colored hind wings (9 genotypes; both genes dominantly control the phenotype), a yellow morph with colored hind wings (3 genotypes; one gene dominant, the other homozygous recessive), a red morph with black hind wings (3 genotypes; one gene dominant, the other homozygous recessive) and a yellow morph with black hind wings (1 genotype; both genes are homozygous recessive) (Ebert *et al.*, 1994).

The authors who first described and named these different morphs of *Z. ephialtes* considered them all to be distinct species. They assigned them different scientific names, with the genus name *Sphinx*: *Sphinx peucedani*, *S. athamanthae*, *S. coronillae*, and so on. This example should be a warning for those making taxonomic classifications according to trait differences. Through the comparison of traits alone, such mistakes cannot be rectified. The existence of morphs demonstrates again how important it is to be skeptical of the use of differences in traits as the only criterion for species membership. Inferring species membership from trait inequality can quickly lead to false results.

Apparently, the two allelic pairs of both of the genes of *Z. ephialtes* that are responsible for the morphs are especially stable. Generally, multiple alleles in the population are not especially long-lived. Selection and genetic drift in most cases ensure that mutant alleles cannot establish themselves in the population. If these alleles have already spread, then they do not often survive much longer. It is an entirely different matter, however, if selection favors the synchronous presence of several variants of an allele, that is, if selection is "interested" in several allelic forms coexisting in the population (see above).

This must be the case with *Z. ephialtes*. It should be noted that the four different morphs are not represented in the expected 9: 3: 3: 1 frequency distribution in nature. Therefore, particular alleles are selected for. However, how are disadvantaged alleles not eradicated by selection? The allelic frequencies differ between the populations of *Z. ephialtes* in different geographic regions. The colored-winged morph primarily populates Central and Eastern Europe, whereas the black-winged morph's distribution is generally in the Mediterranean region and in the western Balkans.

The four different *Z. ephialtes* morphs appear to be integrated in different mimicry systems (Sbordoni *et al.*, 1997). The yellow morph with black, non-colored hind wings exhibits a striking similarity to the nine-spotted moth (*Syntomis phegea*). *S. phegea* belongs to a completely different family, namely the Arctiides (wooly moths) rather than the Zygaenides (burnet moths). *Z. ephialtes* exhibits an impressive example of both mimicry polymorphism and of an astonishing convergence in the appearance of unrelated species. The resemblance of the yellow, black-winged *Z. ephialtes* morph to the Nine-Spotted Moth *S. phegea* is amazing (Color Plate 4).

Yet, this example of mimicry only provides an explanation for one of the four *Z. ephialtes* morphs. Incidentally, there are also locations where both *S. phegea* and *Z. ephialtes* are found and where the representative morph of *Z. ephialtes* is not the mimic. With respect to the entirety of *Z. ephialtes* morphs and their specific frequency distribution in different geographical locations, there is currently no conclusive connection between the respective color patterns and a corresponding mimicry system. Which animal species is mimicked at different locations and thus selectively influences the frequency distribution of the alleles is to a great extent unknown.

5.14 The Color Pattern Polymorphism of the Shells of the Brown-Lipped Snail *Cepaea nemoralis*

What is the selective advantage of intraspecific polymorphisms? This question arises for many animals. Historically, polymorphisms have been treated as neutral, without functional relevance. This position must today be considered incorrect (Jones, Leith, and Rawlings, 1977). Instead, intraspecific diversity is viewed as ranking among the key principles of the survival of certain animal species. An intraspecific polymorphism appears to be one of the best ways to ensure a species' long evolutionary life. It also appears to be incorrect to consider a multitude of types within the species as the starting point for an eventual speciation.

Excellent examples of intraspecific polymorphisms are the color patterns of the shells of the Brown-lipped Snail (*Cepaea nemoralis*). These shells exhibit an extreme diversity of coloration and banding. In the same habitat, the shells of this species can be yellow, red or brown. The shells can also be without bands or display anywhere from one to five bands. There are 20–30 different morphs within a single population. What are the reasons for this diversity? Why is the best-adapted type not dominant? It cannot be the case that all of the morphs, with their differing colorations and patterns, have precisely the same chances for survival and reproduction. Why does one genotype not replace the other? Visual selection by predators or climatic selection should certainly repeatedly prefer one or another phenotype. And if this represents a case of a variable adaptation to different ecological niches, why then does the differing niching not lead to assortative mating and thus to sympatric speciation (Chapter 6).

Whether *C. nemoralis* exhibits bands may have something to do with the likely primary predators of the Brown-lipped Snail in several areas: song thrushes (*Turdus philomelos*) (Jones, Leith, and Rawlings, 1977). In this regard, many investigations have been performed to determine whether the extent of banding or the coloration of the snail's shell has beneficial camouflage effects. Indeed, it was found that song thrushes have more difficulty spotting banded snails in the vegetation than they do spotting non-banded snails. This result does not, however, explain the polymorphisms observed in this species of snails. On the contrary, the polymorphism should eventually disappear, at least in those locations where the song thrush is the primary predator of the snail.

Climate factors may exert a selective influence on the presence and color of bands in *C. nemoralis*. Dark shells absorb more thermal radiation, whereas yellow shells reflect light more strongly. The color of the shell therefore influences the temperature of the snails. Exposure to the sun is crucial. Too much sun leads to heat death, but too little sun leads to the cold-blooded snails being too stiff and having limited mobility, which in turn puts the animals at risk for starvation. However, this explanation is also dissatisfying. If exposure to sunlight is the primary factor in the persistence of the multiple morphs, why are so many different morphs found at the same location?

C. nemoralis lives in a large geographical range in Europe and is found from Norway to Spain and from the coast up to an elevation of 1200 meters. One can observe *C. nemoralis* in habitats as diverse as dunes, meadows and forests. Admittedly, the morphs are not uniformly distributed. Instead, some regions are dominated by certain shell variations. However, it is not the case that every geographical location, every mountain peak or every habitat hosts its own specific morphs. Furthermore, it has been observed that different morphs remain in the sun for differing time periods (Jones, Leith, and Rawlings, 1977). Some morphs prefer the cool morning hours for their activities, others are active around noon, and still others are predominantly active in shadows. However, no conclusive connection between the ambient temperature and the snails' coloration or degree of banding has been observed. In most cases, scientists cannot assign the specific color patterns to the individual climatic factors of the morphs' respective habitats; the situation is too complex.

The driving force for the intraspecific diversity of *Cepaea* snails appears to be that increased species diversity leads to an increase in the number of potentially habitable environments. By increasing their own diversity, the snails expand their resources and thus increase their population size. This would without doubt be a selective advantage. However, the question of why snails living in different niches have not speciated remains unanswered. This question likely ranks among the crucial questions in the field of taxonomy.

5.15 The Beak Polymorphism in the Black-Bellied Seedcracker Finch *Pyrenestes ostrinus*

The beak size of finches is strongly controlled by selection. Given that beaks can be used for cracking certain seeds for food procurement, fluctuations in beak size as small as a tenth of a millimeter can lead to the affected birds no longer being competitive against their conspecifics. The size of the beak, whether it is slightly smaller or bigger, is not the result of special training by the bird; rather, this trait is genetically predetermined (Abzhanov *et al.*, 2004).

The finch's beak offers a fine example of a stable polymorphism within a population, similar to blood groups among humans, as discussed above. The West African Black-bellied Seedcracker (*Pyrenestes ostrinus*) provides a textbook example of adaptive polymorphisms, which refers to the coexistence of different adaptations to different habitats by representatives of the same species. In Cameroon, two morphs of the Black-bellied Seedcracker are characterized by different beak sizes, and both morphs occur simultaneously in a single population (Smith, 1993; Smith, 2008). The small-beaked individuals primarily feed on soft seeds, whereas the large-beaked birds are specialized in cracking hard seeds. This is a case of dimorphism.

This dimorphism is comparable to sexual dimorphism. Only thin-beaked and thick-beaked morphs exist; no animals with intermediate beak sizes are observed, just as no intermediate sexes are observed. However, there is a significant difference between these two examples of dimorphism. In the case of sexual dimorphism, the sole occurrence of two distinct types is because of a genetic mechanism that determines the either-or result. With respect to the finch's beak, however, selection is responsible for the fact that no mixed types are observed. Individuals with beaks of intermediate size admittedly develop due to various allelic combinations, but because of selection's severe control, these animals have no chance for survival compared to their "pure" small- or large-beaked conspecifics that are specialized to eat very particular seeds. All of the intermediate phenotypes are continuously weeded out. Selection divides the population into only two morphs. In the case of these finches, there is neither a genetic mechanism for the development of only two beak sizes nor any mating barriers between the two morphs with the two beak sizes. Both of the morphs belong the same species given that they are members of the same gene-flow community.

The simultaneous coexistence of two phenotypes within the same population makes biological sense. However, it is not the selective advantage of the thick or thin beak that is at work here. Instead, it is the clear selective advantage of there being two variations that together enrich the entire population. The existence of two morphs with differing beak types broadens the food spectrum of the species. If all of the birds had the same beak type, then they would all eat the same food. Accordingly, the food resources in the habitat would become depleted much more rapidly.

The example of the African Black-bellied Seedcracker is similar that of certain raptors, namely that the females are substantially larger than the males. This is, for example, the case with goshawks, sparrow hawks and peregrines. Apparently, the dimorphism of two body sizes expands the available food spectrum. To some extent, the larger females hunt different prey than the smaller males. The sexes compete less with one other and utilize a larger range of prey.

Again the unsolved taxonomical problem arises of why nature pursues two different paths to conserve novel adaptations. One way is sympatric speciation; the other way is adaptive polymorphism within the species.

Many examples illustrate that genetic adaptations that make an additional food source accessible lead to sympatric speciation (Dieckmann *et al.*, 2005; Tautz, 2009). If there is a selective pressure to preserve adaptive niching in a new habitat, then there is also selective pressure to prevent backcrossing with their conspecifics. The coexistence of adaptive niching and assortative mating leads to sympatric speciation. Yet, why is there a second way that preserves novel adaptations to new habitats? Why do intraspecific stable polymorphisms occur as alternative to sympatric speciation?

5.16 The Beak Polymorphism in the Darwin Finch *Geospiza fortis*

An additional example of an intraspecific polymorphism that apparently increases the flexibility of environmental adaptations is the beak of the Darwin Finch *Geospiza fortis* (or the Medium Ground Finch). Similar to the African Black-bellied Seedcracker, the ground finch populations on the Galapagos island Daphne Major consist of strong-beaked and slim-beaked individuals. In contrast to the Black-bellied Seedcracker, however, the strong- and slim-beaked ground finches do not coexist but succeed each other in different years according to climatic conditions. In a 30-year study, the Grant couple examined how the prevalence of the morphs alternated year by year (Grant and Grant, 2002).

On the island Daphne Major, periodic variations of the sea current (El Niño) cause a dry period that lasts three to five years. This period is then followed by a wet period that lasts equally long. A result of this cycle is that the available vegetation also changes rhythmically. In wet years, the finches have smaller, softer seeds available; in dry periods, plants with thick, hard seeds prevail. This phenomenon is accompanied by a remarkable oscillation in the beak size of the Medium Ground Finch *Geospiza fortis*. These finches have slightly slimmer beaks in the wet years and thicker beaks in the dry years. This is by no means due to training effects but rather to allelic differences (Abzhanov *et al.*, 2004). The gene pool of the entire population of Medium Ground Finches contains the alleles for both thick and thin beaks. The only factor that changes during the particular periods is the allelic frequency distribution among the living individuals.

As soon as a dry period begins, the majority of the Medium Ground Finches starve due to the absence of soft seeds. It is primarily the thick-beaked individuals that survive because they are able to crack extremely hard seeds. For a few generations, these surviving birds produce descendants with larger and broader beaks. After a few years, however, the wet weather returns, allowing soft-seeded plants grow again. The opposite situation then faces the finch population: most of the now common thick-beaked finches die and the few surviving thin-beaked finches immediately have a selective advantage, produce higher numbers of descendants and begin to displace the thick-beaked individuals.

The *Geospiza fortis* is an impressive example of a stable allelic polymorphism. A strong selective pressure influences the respective composition of the allelic distribution in the population without completely displacing one variant of the alleles. The two morphs alternate in their dominance over very few generations due only to differences in allelic frequencies. The extinction of the population is thus avoided. In cases where the environment changes over time, the population is prepared for this change. That is, the population always maintains a few ready individuals that are preadaptively adjusted to a possible climate change. These individuals can replace the newly disadvantaged organisms in a relatively short period. A remarkable fact is that neither those alleles for developing thin nor thick beaks become completely extinct, although such beaks are disadvantages during certain periods.

This phenomenon is difficult to understand Darwinistically. Given that thin beaks are disadvantageous during dry periods, the alleles for developing thin beaks should disappear in these times. Similarly, the alleles for developing thick beaks should disappear in the wet periods. Yet, neither of these alleles disappear due to negative selection or genetic drift. This is then a case of a stable polymorphism. The selective advantage of a stable polymorphism is not the advantage of one or the other allele. It is the advantage of versatility, of preadaptation to conditions that currently do not prevail but may in the future. If the environmental conditions change, the populations have the advantageous alleles readily available. This concept is related to group selection, the controversy of which has been discussed in many papers (Okasha, 2001; Okasha, 2006).

In recent years, there has been important progress in understanding the genetic control of the finch's beak (Abzhanov *et al.*, 2004). The morphogenesis of the bird's beak is indeed controlled by one or a small number of genes, and key among these genes is Bmp4. This structural gene has far-reaching tasks in the differentiation of the cells that originate in the cranial neural ridge and, accordingly, cranial skeletal morphogenesis. Bmp4 controls the patterning, growth and chondrogenesis of the mandibular and maxillary protrusion of the anterior head skeleton. As is common with regard to animal morphogenesis, this gene plays different roles during different developmental periods.

The zoologist Arkhat Abzhanov, a native of Kazakhstan and currently conducting research at Harvard, has examined six species of Darwin's Finches with respect to species-specific differences in Bmp4 expression. Differences with respect to the time and extent of Bmp4 expression correlate with differences in the beak morphology of the different Darwin's Finches. Underlying these differences are, predictably, regulatory DNA sequences (enhancers) and tissue-specific transcription factors that are species-specifically expressed in the developing beak epithelium. Therefore, the allelic differences among the beak-morphs of the finches cannot be observed in the structural gene Bmp4 itself but in its regulatory elements; this structural gene is variably regulated but is otherwise unaffected. Experimental interventions that artificially increase the level of Bmp4 expression have indeed resulted in alterations of beak morphology.

5.17
Intraspecies Morphs in the Garter Snake *Thamnophis ordinoides*

In the case of the North American Garter Snake (*Thamnophis ordinoides*), several morphs coexist in the same population (Brodie, 1989). These morphs differ significantly with regard to their color pattern and behavior. The different morphs also vary in having large or small hunting grounds. In the face of danger, some morphs flee, while others stay motionless or exhibit aggressive behavior. It is interesting that the different traits (i.e., pattern, color, and behavior) are correctly tuned with one other. Snakes that display camouflage-colored dorsal patterns also exhibit motionless behavior when in danger; in contrast, snakes that display more a striking coloration behave aggressively or flee.

The genetic foundation for the respective coexistence of these traits is the linkage of their encoding genes. The genes for the different traits (i.e., pattern, color and behavior) are located on a single chromosome, where they form a linkage group that is protected from recombination. This arrangement leads to clearly identifiable intraspecific morphs and to a situation in which a particular color pattern is combined with a particular antipredator behavior.

Again, the selective advantage of the existence of morphs is that multiple adaptations are present within the same species. In the example of the Garter Snake, in contrast to the Black-bellied Seedcracker finch, however, the multiple adaptations are genetically controlled, not environmentally by selection.

5.18
Urbanization in Certain Bird Species is based on Genetic Polymorphism

Preadaptive genetic dispositions often appear to be responsible for behavioral changes; training or learning processes are not always involved. An increasing number of bird species has moved into the cities over the last centuries, particularly during the previous few decades. There are examples supporting the view that urbanization is based on genetic preadaptation.

The timidity of many animals to both humans and certain predators is the result of experience and genetic dispositions. Imprinting during early life is crucial, as are later experiences and education by parents and other conspecifics. However, a great deal of timidity toward potential predators is genetically founded. In the case of birds, it appears that the timidity toward humans differs widely between geographical regions and has changed in time.

Although this observation first appears to support the position that timidity is a result of varying experiences with humans, many data also support a predetermined genetic disposition. A newly hatched chicken can distinguish between a sparrow and a harmless dove by their flight silhouettes without having learned of this difference from their parents or other conspecifics. A sandpiper (*Calidris* spec.) from northern Siberia that arrives in the Wadden Sea off the German North Sea coast in August and which was born during the same summer can distinguish a human or a dog from a sheep grazing on a dyke without having had previous experiences with humans, dogs or sheep.

Advocates of the position that birds' timidity toward humans is solely based on negative experiences do not have arguments for why the European White Stork (*Ciconia ciconia*), within living memory, has been fearless of humans, while its close relative, the Black Stork (*Ciconia nigra*), is extremely fearful of humans. In several breeding areas, both species are not persecuted, and persecution on the migration routes and in the hibernation regions is not different between White Storks and Black Storks.

When the cities that were built over the past century or more began covering increasingly large areas, surpassing rural areas and forests with food availability and protection against predators, several bird species moved from their ancestral habitats. Birds moved into cities to an increasing extent, where they found a larger food supply and better protection from predators. Examples of species that moved to the cities include the Grey Heron (*Ardea cinerea*), several duck species, the Coot (*Fulica atra*), the Moorhen (*Gallinula chloropus*), the Woodpigeon (*Columba palumbus*), the Mistle Thrush (*Turdus viscivorus*) and the Carrion Crow (*Corvus corone*). Remarkably, the Woodpigeon still maintains its name from past times, although it has conquered the cities and urban regions in most of Europe and no longer primarily occupies forests. The urbanization of the Woodpigeon occurred in northwest Germany in the 1950s and 1960s. Before this time, this pigeon was a shy forest inhabitant. Among hunters, the saying "this is a woodpigeon" signified a special honor, as a successful shooting of a Woodpigeon signified rarity and the hunter's ability. Today, Woodpigeons nearly feed out of one's hand.

Yet, why did it take so long for pigeons to conquer cities? The lesser danger of being shot in cities and a larger food supply was certainly available for a much longer time than the last half century. Even today, there are regions in Europe in which the Woodpigeon is not urbanized. For example, such is the case in south and southeast central Europe, where hardly any urban Woodpigeons can be found (Bezzel and Kooiker, 2003).

The geographical differences in the timing of urbanization of different bird populations illustrates that this process cannot be solely based on learning processes. Therefore, timidity toward humans is most likely based in part on a genetic disposition, and this disposition appears to be polymorphically anchored in a population. A genetic disposition to fearlessness would explain why this behavior arose in different places at different times. An increase or decrease in timidity towards humans is subject to a selective advantage. Consequently, a single or small number of alleles can displace other alleles among a population. The clear selective advantage of being amenable to cities, which are rich in food and safety, has led to significant changes in allelic frequency distributions in several bird species in the course of only a few generations.

5.19
The Mimicry Morphs of the Female Swallowtails of the Genus *Papilio*

Intraspecific morphs were of particular concern to taxonomists when it came to changing taxonomy from a purely trait-oriented grouping process to one that was more scientific. More than one hundred years ago, Edward Bagnall Poulton recognized the inconsistency and contradictory nature of a species concept based only on traits. The importance of his classical publications "What is a species?" and "The conception of species as interbreeding communities," published in 1904 and 1938 in the "Proceedings of the Entomological Society of London" and the "Proceedings of the Linnaean Society of London" respectively (see Mallet, 2004), can hardly be underestimated. These publications stripped the Linnaean typology from its scientific foundation and exposed the classification by traits to be a subjective grouping according to our own pragmatic objectives and trait assessments, that is, that species were, as Darwin said in 1859, "made for our convenience" (Darwin, 1859). Moreover, Poulton anticipated Ernst Mayr's essential lines of thought by decades, proposing the fundamental concept of the species as a reproductive community.

Poulton had been influenced by the observations by Alfred Russel Wallace and Henry Walter Bates of mimicry morphs in butterflies (Mallet, 2004). Bates primarily investigated South American Heliconiids, whereas Wallace discovered the fascinating mimicry morphs of the Southeast Asian Swallowtails *Papilio memnon, P. polytes* and *P. aegeus*. There are several different morphs among the females of these butterfly species. In contrast, there are no morphs among the males, which have uniform shapes. The different female morphs mimic butterfly species of entirely different families, which are inedible to birds. The caterpillars of the mimicked species consume poisonous substances that are still present in the body of the imaginal stage butterfly. The mimics do not contain any unpalatable substances, but due to their similar appearance, the imitators (just as the imitatees) are protected from predators. This has an immense selective advantage and is a classic case (and historically one of the first published examples) of Batesian mimicry, referring to the imitation of a dangerous or inedible animal by a harmless animal to mislead the latter's predators (Wickler, 1968).

For Poulton, the importance of these butterflies lay most of all in the fact that erroneous conclusions could be drawn when phenotypes were used to determine species affiliation. The females of swallowtail butterflies can differ much more strongly from members of their own species than they do from the members of other swallowtail species. The same phenomenon was later observed in the African Swallowtail *Papilio dardanus*, which is today counted among the best known examples of Batesian mimicry (Mallet, 1995). Here, too, the males are uniformly shaped, whereas the females are polymorphic. In both Madagascar and Ethiopia, only original monomorphic, non-mimicking populations are found. In these groups, the females, like the males, exhibit distinctive tails on their hind wings and a yellow-black coloration, hence appearing to be a common *Papilio* butterfly. In other locations in East Africa, the females of *P. dardanus* mimic the inedible butterflies of other families, for example, the milkweed butterflies *Danaus chrysippus* and *Amauris niavius*, which belong to the family *Danaidae*. Remarkably, only female milkweed butterflies exhibit this polymorphism, whereas the males are uniformly shaped. In total, more than 30 significantly distinguishable mimetic female *P. dardanus* morphs have been discovered (Salvato, 1997). In certain *P. dardanus* populations, the females only mimic one other species; in other populations, however, several species are mimicked in the same location.

This form of mimicry polymorphism is not restricted to the genus *Papilio*. An analogous example can be observed in another butterfly family, the Nymphalides. Here, the species *Hypolimnas misippus* exhibits a similar polymorphism, which is also restricted to the female of the species. The males are uniformly shaped, whereas there are several morphs among the females. Among these morphs are those that again imitate *Danaus chrysippus*. *Hypolimnas misippus* is found across Africa, Asia and Australia.

The genetic foundation for this sex-linked polymorphism is apparently based on a complex of linked genes that are protected from crossing-over events. Each morph expresses its own linked group of genes, resulting in a well-adjusted combination of color pattern, wing shape and behavior. Intermediate types do not exist, at least not in noticeable numbers. Instead, there are only clearly distinct morphs. In addition to the *Dardanus* morph, there is also the *D. chrysippus* morph, the *A. niavius* morph and several others.

The entire morph-determining gene complex appears to be located on the Y chromosome, which, because the Y-chromosome is largely protected from recombination, explains the linkage of the genes. In the case of butterflies (in contrast to most other animals), the females are the heterogametic sex. Accordingly, they are of the XY constitution, as are male mammals, rather than the XX constitution. Generally, the sex chromosomes of heterogametic females are not designated XY, but ZW. In any case, the Y (or W) chromosome is largely protected from genetic recombination, such that the genes of the morphotype complex cannot be separated. The linkage of the morphotype complex on the Y chromosome is a plausible explanation for the exclusive female-linked occurrence of mimicking morphs among butterflies.

Historically, the conspecificity of the different *Papilio* morphs was discovered in the wild by observing the copulation of different female morphs with affiliated males. The occurrence of such different morphs that belong to the same sexual community induced Poulton to propose new species concepts. He refused to purely sort organisms into diagnosable units and replaced the typological species concept with a concept that was based on natural laws. He proposed to unify mutually reproducing organisms into a species, referring to species under this new concept as "syngamic species." Thus, he anticipated Mayr's species concept of the reproductive community (Mayr, 1942) (Chapter 6). Introducing an additional species concept, Poulton also combined the descendants of common ancestors into a group; he referred to species under this classification as "synepigonic species." This was, roughly speaking, the anticipation of the cladistic species concept (Hennig, 1966) (Chapter 7).

5.20
The Morphs of the Brood-Parasitic Cuckoo Female *Cuculus canorus*

During the breeding season, the Eurasian Cuckoo (*Cuculus canorus*) ranges throughout Europe and as far as Eastern Asia. Its most striking trait is brood parasitism. The cuckoo is unable to build nests, to incubate its eggs or to raise its young. Instead, it lays its eggs in the nests of alien bird species, with these being nearly exclusively members of the suborder *Passerines* (*Passeriformes*). These host birds generally do not recognize the fake brood. Rather, the host bird incubates the cuckoo's egg and feeds the fledgling until it is fully fledged and no longer depends on the foster parent's care.

The cuckoo is extraordinarily relevant to the discussion of the species concept, as every cuckoo female is genetically adapted to a specific host bird with respect to a large number of traits. Every cuckoo female has only a single species of host birds in whose nests it lays its eggs. A robin cuckoo lays its eggs only in the nests of robins. The cuckoo female does not change hosts, even if it cannot locate any nests of its specific host during the egg deposition period. The female offspring of a robin cuckoo lays its eggs only in the nests of robins again in the following spring.

Just in Europe alone, there are approximately 90 different species of host birds to which cuckoo females are adapted (Glutz von Blotzheim, 1994). Accordingly, there are 90 genetically different cuckoo populations. However, not all species of host birds are parasitized by cuckoos with the same frequency. Some host species are visited by cuckoo females especially frequently; other species are visited by only a few cuckoo females. This is an example of the equal distribution of the alleles in a polymorphic population. Moreover, there are still other species of host birds for which the cuckoo offspring has only a small chance to fledge successfully. There are many additional species of host birds in Asia, bringing the total number of cuckoo host species and cuckoo populations to over 200 (del Hoyo, Elliott, and Sargatal, 1997).

The strong adaptation to only one species of host birds is determined by numerous heritable traits. The cuckoo female must be able to lay its eggs at the correct time. European Robins (*Erithacus rubecula*), Black Redstarts (*Phoenicurus ochruros*) and thrushes (*Turdus* spec.) are frequently visited hosts in Central Europe and breed at the end of April. In contrast, reed warblers (*Acrocephalus* spec.) breed only at the end of May. The cuckoo must be hormonally adapted to very different egg deposition times; its eggs must be ready for deposition neither too early nor too late.

The cuckoo must locate and recognize the nest of the correct host bird. To this end, it can at best rely only on memories from the previous year, when it was raised in such a nest. However, except for certain reminders of its own place of birth, it of course never received parental instruction regarding (1) how to locate a suitable breeding biotope, (2) locating a potential host bird's nest, be it in a tree, in low shrubbery or on the ground, or (3) distinguishing such a nest from the nests of other hosts. Cuckoo females locate suitable nests by observing the host bird preparing its nest and beginning incubation, waiting for hours in a concealed place if necessary. Moreover, it is probable that the host birds are identified acoustically by their song. The ability to locate a proper nest must also be determined to a large extent hereditarily given that no parental instruction is possible and that the only available memories are from their own hatching and raising.

Furthermore, brood parasitism can only be successful when the cuckoo's eggs mimic the eggs of the host birds with respect to size, color and pattern. In the Eurasian Cuckoo *Cuculus canorus*, there are more than a hundred genetic variations that determine the eggs' sizes and colorations. The individual female, including its female descendants, only lays eggs of one particular type. This egg mimicry can only be explained by the presence of polymorphic alleles.

With respect to understanding the species concept, the primary challenge is not simply explaining perfect adaptation but how this adaptation can be explained by a number of linked genes, all of which must work together as a linkage group in a cuckoo female. The time of egg deposition, the identification of the host and the coloration of the eggs are supported by genetic dispositions that must not be separated by genetic crossing-over events. This is because a particular cuckoo female cannot have half of its traits adapted to one host bird and the other half to another host bird. None of the adaptation traits that are attuned to a particular host bird can recombine. The sum of these traits must be inseparably and genetically linked to each other. Accordingly, there are genetically distinguishable Robin cuckoos, Black Redstart cuckoos, Thrush cuckoos, Reed Warbler cuckoos, and so on. Why, then, are not there 200–300 species of cuckoos in Eurasia?

The answer to this question is as follows. The adaptation to a particular species of host bird is only maintained by the female cuckoos. In males, there is no host specificity. Accordingly, just as with the mimicry polymorphism of female Swallow-tail butterflies (see above), the cuckoo morphs are sex-related.

Why is it the female sex that exhibits polymorphism? It does not appear to be a coincidence that both butterfly and bird morphs are determined exclusively by the female sex. This is because in birds, just as in butterflies, females, as an exception, are the heterogametic sex. In butterflies and birds, the females are of the XY constitution (occasionally referred to as ZW), whereas the males are of the homogametic XX constitution. The Y (or W) chromosome contains a large percentage of genes that are embedded in heterochromatin and thus are protected from recombination. These protected genes are referred to, in this instance, as a linkage group. It is a safe assumption that all of the genes that determine the distinct phenotype of a particular morph are located on the Y chromosome, the genes of which are generally protected from genetic crossing-over. This situation would explain (1) how an entire battery of genes is linked and inherited as a whole and (2) why these genes are inherited along the maternal line. If the cuckoo was homogametic in the female sex, as in mammals, brood parasitism likely could not exist.

6
Biological Species as a Gene-Flow Community

6.1
The Definition of the Gene-Flow Community

The species concept of the gene-flow community has been established at the beginning of the twentieth century by the British scientist Edward Bagnall Poulton (Poulton, 1903; Poulton, 1938). It was then in 1937 substantially upgraded by Theodosius Dobzhansky, a geneticist who was born in the former Soviet Union, by combining taxonomy with genetics (Dobzhansky: "Genetics and the origin of species" 1937). In 1942 the concept received a successful advocate in Ernst Mayr who propagated this species concept for a broad audience (Mayr: "Systematics and the origin of species" 1942). Finally, Michael Ghiselin then substantiated this species concept philosophically in 1997 (Ghiselin: "Metaphysics and the origin of species" 1997). Despite its many fathers, the concept of the gene-flow community is today most of all associated with the name Ernst Mayr; for it was him who popularized this concept and referred to it as the actual "biological species concept," so that Mayr is displayed in many textbooks as the originator of this species concept.

A gene-flow community is a community of organisms that are connected by gene flow through sexual reproduction. Sexual gene flow is also called lateral gene flow and is opposed to vertical gene flow, which is the transfer of genes from parental organism to its filial generations. Interorganismic connection by lateral gene flow is a distinctive feature only for those organisms that have a biparental reproduction, (organisms that can only reproduce if there are two parents). Plants and animals that undergo only uniparental reproduction are clearly distinguished from biparental organisms because they are not connected by lateral gene flow. Those organisms that reproduce only vegetatively or parthenogenetically have no sexual gene transfer and therefore cannot be considered to form gene-flow communities. As the species concept of a gene-flow community cannot be applied to uniparental organisms, all of those organisms are, in the sense of a gene-flow community, "species-less" (Ghiselin, 1997). If the species definition of lateral gene flow were applied to

Do Species Exist? Principles of Taxonomic Classification, First Edition. Werner Kunz.

uniparental organisms, then each "birth" would be a speciation because, due to the lack of biparental reproduction, no individual will ever recombine its genome with the genome of another. Of course, it is not reasonable to apply the term "species" to an individual.

A particular problem is posed by prokaryotic organisms, as they also have lateral gene transfer, because in its general sense each transfer of genes between two organisms other than a transfer from a mother into a daughter is a lateral gene transfer (see below). Although the biological mechanisms of prokaryotic lateral gene transfer are different from the mechanism of sperm-egg fusion in eukaryotes, the final population-genetic and evolutionary results of both forms of gene transfer are similar. In both cases, organisms that may have a long evolutionary history of separation merge again and exchange genes. In both cases, the result of such contacts is the recombination of genomes. In both cases, lateral gene transfer makes it difficult to generate a dichotomously branching phylogenetic tree. Even distant branches may merge, converting the tree into a connected network (Figure 6.1).

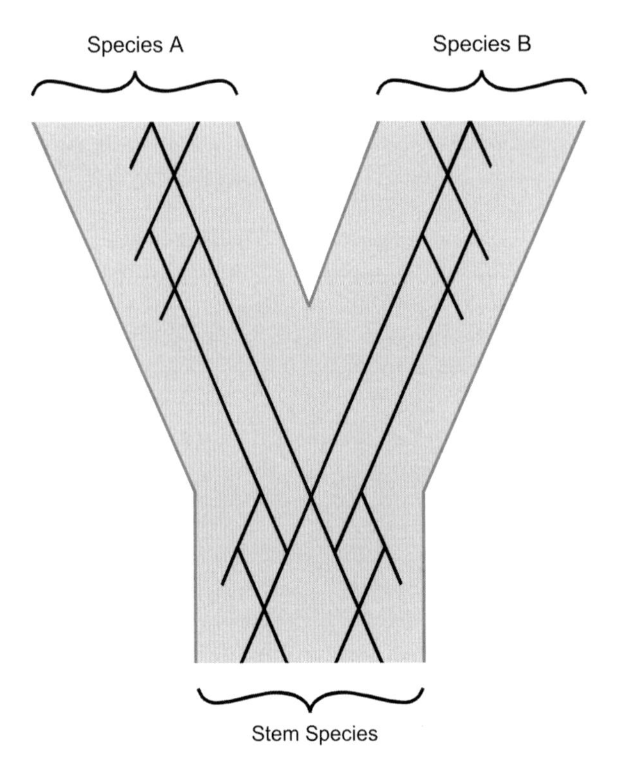

Figure 6.1 In a cladistic tree of biparental organisms, the branches refer to taxa, not to organisms, whereas the reticular cross connections within the shrouded branches refer to individual organisms. A cladistic tree of biparental organisms cannot be displayed at the organismic level; it only can be displayed at the taxon level. This difference between the cladistic trees of bi- and uniparental organisms reflects the reason why uniparental organisms cannot be considered to be species.

The connection of organisms by sexual (lateral) gene flow must be distinguished from the connection of organisms by genealogical (vertical) gene flow. These processes differ, although eukaryotic sexuality in most (but not in all) cases is connected with reproduction. Biparental sexual gene flow in eukaryotes involves the fusion of a sperm and an egg and the recombination of two genomes. Genealogical gene flow is the production of filial generations by the parental generation.

There are biological and logical differences between biparental sexual gene flow and genealogical gene flow:

1) First, the biological difference: Biparental sexual gene flow is the transfer of DNA between two cells with different genomes; in higher eukaryotes, this occurs in the form of a fusion of sperm and egg. Genealogical gene flow, in contrast, is the passage of identical genomes from the mother to the daughter organism via the transformation of a zygote into an embryo.

   Biparental sexual gene flow in eukaryotes always proceeds in two steps. First, the two genomes of the diploid chromosomal set are recombined, and then the genomes of two different organisms merge with each other by the fusion of a male germ cell with a female one. The first process is called meiosis; the second is fertilization. Both processes may be separated from each other by a considerable span of developmental life, for example, in lower plants.

   In the case of metazoans (multicellular animals), these two sexual processes are always followed by reproduction and vertical gene flow into a filial generation, but in principle, the fusion of a male with a female gamete (fertilization) does not necessarily result in reproduction or multiplication (propagation). This phenomenon is nicely documented in Ciliates, in which the sexual merging of two cells that leads to the fusion of different genomes is only temporary (and is called conjugation, not fertilization). After the genomes merge, the two conjugating partner cells separate again and continue as two genetically altered individuals without having produced any sons or daughters. Sexuality results primarily in the acquisition of new genetic material and is only secondarily related to the production of filial generations.

   In mosses, ferns and flowering plants, the decoupling of the process of sperm-egg merging (fertilization) from the process of the production of offspring (reproduction) is significant. When a male germ cell combines with a female germ cell to form a zygote, then no new individual originates from this process. No F1-generation is produced by the parental generation. The plant individual having produced a zygote just keeps growing. The zygote produces the sporophyte tissues, and the sporophyte continues to grow on the gametophyte as a part of the same organism. The sporophyte and gametophyte constitute same individual in mosses, ferns and flowering plants. The sporophyte is not a new individual, but part of an existing plant.

   In summary, nature offers many examples of the decoupling of sexual gene flow and genealogical gene flow. The Ciliates and the plants demonstrate fertilization without reproduction or multiplication. In contrast, vegetative or parthenogenetic propagating organisms demonstrate reproduction or multiplication without fertilization.

2) Now, the logical differences between biparental sexual gene flow and genealogical gene flow: the connection of the organisms via biparental sexual gene flow is symmetrical, whereas the connection of the organisms via genealogical gene flow is non-symmetrical. Sexual gene flow means that the organisms of population A mate with the organisms of population B, and conversely, the organisms of population B mate with the organisms of population A.

   In contrast, the connection of organisms via genealogical gene flow is not symmetrical over successive generations. If organisms pass their DNA on to organisms of the subsequent generation (mother – child – grandchild), then gene flow is not reversible. Genealogical gene flow runs along the temporal axis, and the course of time is not reversible. After birth, the F1 organisms are no longer genetically connected to their creators, in contrast to the symmetry of biparental sexual connections. If, in spite of this constraint, sexual contact between members of the daughter generation and members of the mother generation occurs (as happens frequently in nature), then this event does not contradict the statement that genealogical gene flow is asymmetric. Sexual connections between individuals of the daughter generation and individuals of the mother generation clearly are lateral gene flow, which is unrelated to the vertical gene flow of the genealogical connection.

6.2 The Connection of Organisms in a Gene-Flow Community Includes the Genealogical Connection

The attempt to separate the process of biparental sexual connection via fertilization from the process of genealogical connection via reproduction poses a difficult problem. In monogamous organisms, biparental connection combines only two genomes, not all genomes in a gene-flow community. In polygamous organisms, biparental connection combines more genomes, but far from all of the genomes in a gene-flow community; a contemporarily cohesively connected gene-flow community does not exist.

The entirety of the group of the gene-flow community cannot be defined by bisexual connections alone. The bonds of a gene-flow community must include genealogical connections.

Not all organisms of the parental generation (P generation) are connected by lateral gene flow. To complete the network of lateral connections, one has to include the successive generations. If two lineages are not laterally connected in the P generation, they will be connected in one of the following F generations. Their children, grandchildren or great-grandchildren will build the lateral sexual bridges among the genealogical lineages. The sons or daughters of the parental generation will find their sexual partners among the sons or daughters of other parents, and this combination of genealogical and biparental sexual gene flow continues along the further F generations, when the grandchildren or great-grandchildren form lateral sexual connections with the grandchildren or great-grandchildren of still other lineages (Figure 2.7).

Thus, the gene-flow community is a network of biparental sexual connections and genealogical connections. To find all the lateral (horizontal) connections, one must step down many generations. Finally, however, all members of the gene-flow community are connected within this network. An important component of the gene-flow community is the repeated lateral combinations of genomes after only a few or hundreds or even thousands of generations. The genes "flow" among the organisms of this network, connecting the members of the gene-flow community. This criterion differentiates the species as a gene-flow community from a group of organisms that propagates only by uniparental reproduction where there is no gene exchange between lineages.

6.3
The Species is a Gene-Flow Community, Not a Reproductive Community

It would be of particular interest to follow individual alleles within a gene-flow community, but the available scientific information is insufficient. For example, the Willow Tit (*Parus montanus*) is distributed continuously from Western Europe to Eastern Asia. Imagine that a single allele mutates at a certain moment. How many generations would it take for a newly mutated individual allele to migrate from Western Europe to Eastern Asia? Would this migration ever be possible (see below)?

Almost no information is available to answer this question. The question becomes even more difficult if one considers that most alleles are short-lived. Individual alleles disappear by selection or genetic drift (Chapter 5); however, to understand the connection among the individuals of a gene-flow community, it is sufficient to consider only gene flow among immediate neighbors. Race *A* is connected with its neighbor, race *B*, and race *B* with its immediate neighbor, race *C*, and so on (Figure 6.2a).

A hypothetical structure of the gene-flow connection is illustrated in Figure 2.7. In many cases, the connection includes only the members of the geographically adjacent race. Thus, a single newly mutated allele (1) in race *A* reaches only the organisms of the nearby resident race *B*, not the members of the more distant race *C*. In turn, a single newly mutated allele (2) in race *B* reaches only the organisms of the nearby resident race *A*, and a newly mutated allele in race *C* (allele 3) reaches the organisms in adjacent race *B*, but not the distant organisms in race *A*. Nevertheless, all races *A*, *B* and *C* are connected like the links of a chain, without direct contact among all links.

Many alleles that arise in race *A* may not arrive in race *C*, and vice versa, at least not across large geographic distances and within the evolutionary life span of an average species. In these cases, the gene-flow connection is non-transitive; from the concurrent connections of *A* and *B* and *B* and *C*, no connection between *A* and *C* can be inferred. Although lateral gene exchange occurs between *A* and *B* and also between *B* and *C*, it may not necessarily occur between *A* and *C* (Figure 6.2a).

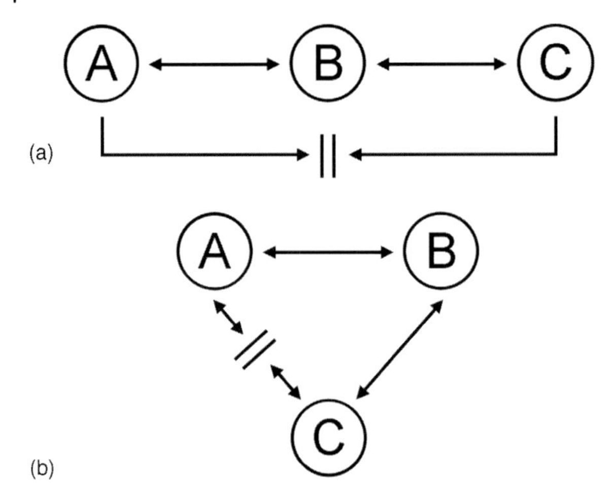

Figure 6.2 Isolation by distance (a) and ring species (b). To understand the cohesion among the individuals of a gene flow community, it is sufficient to consider only the immediate neighbors. Race *A* is connected with *B*, and both races exchange alleles. The same is true for *B* and *C*, but the geographically distant races *C* and *A* do not necessarily exchange alleles and may not necessarily be cross-fertile. Many alleles that arose in *A* may never be observed in *C* and vice versa. A ring species (b) is not a taxonomic peculiarity, not to mention a kind of "super-species." It is nothing more than the frequently observed phenomenon of isolation by distance, distinguished only by the peculiarity that distant races encounter each other again under natural conditions.

Reduction in gene flow has consequences. Genetic compatibility between distant races of a species may become reduced. Although studies on this subject are few, it is likely that the crossing of distant individuals of a species in many cases does not result in completely viable and fertile offspring, at least in the case of large geographical distances (Ford, 1954; Irwin *et al.*, 2005). This means that distant organisms of a species do not belong to a "potential reproductive community." A "potential reproductive community" is made up of distant or isolated groups of individuals that could successfully reproduce together if they were to encounter each other (Mayr, 2000) (see below).

For this reason, the traditional term "reproductive community" (Mayr, 2000) does not reflect the nature of the group connection as it is defined in this book; the term "reproductive community" conveys the impression that all members of this community reproduce with each other, at least in the course of evolutionary time. This assumption, however, often does not hold. In some animals and plants, geographically distant organisms of a species never exchange alleles, even though they are considered to belong to the same gene-flow community. However, distant organisms of a species are still connected by uninterrupted gene flow, even when they have lost their reproductive compatibility (Figure 2.7). For this reason, in this book, the more precise term "gene-flow community" replaces the conventional term "reproductive community" (Mayr, 2000).

The logic behind this consideration is as follows: local adaptation requires the alteration of many different traits that are genetically determined. These genes must function concertedly in the organism. If two organisms with very different adaptive traits crossbreed with each other, the genes of these adaptive traits will lose their chromosomal linkage when they recombine. Consequently, the offspring of such crosses should be intermediate with respect to their adaptive traits. It is unlikely that these hybrids would be fit enough to compete successfully with pure-bred organisms, at least not in those cases where the local adaptations involve an array of cooperating genes that would be disrupted by recombination.

Despite this loss of reproductive connection, however, such distant populations are connected via stepwise gene flow. They are joined via "intermediate organisms" in the connecting regions (regions of clinal transition). Thus, all organisms form a common gene-flow community, even if individual alleles do not migrate over the entire distribution area of the species. The gene-flow community also includes distant populations whose members have even lost their potential reproductive compatibility. The distant organisms in a gene-flow community are only incrementally connected to each other; for example, a human's foot is connected to his head, without the foot being in direct contact with the head.

6.4 A Species Concept Requires Both Connection and Delimitation

Taxonomy is the science of group connection; however, groups are not only defined by connection of the conspecific individuals but also by delimitation against unrelated individuals. Those characteristics that connect organisms as groups and separate them from other organisms define a natural taxon (Mishler and Brandon, 1987; Ereshefsky, 1999). In contrast to most other species concepts, the concept of the gene-flow community satisfies both conditions. Gene-flow connection includes both connection and delimitation. The organisms of a gene-flow community are joined by the exchange of genes and delimited by the barriers to this gene flow. Most other species concepts are incomplete, because they do not contain the criteria of delimitation.

The barriers between two gene-flow communities inhibit gene flow and prevent (or allow only to a limited extent) gene flow from one species into another. Barriers such as these are natural boundaries between two groups, not human-made delimitations. This point is not changed by the discovery that gene-flow communities are, in many cases, leaky. The fact that boundaries are vague does not contradict the fact that boundaries exist. A partially permeable boundary is still a boundary (Figure 2.7).

Several other species concepts include only the criteria for inclusion; they do not tell us by what criteria one species is delimited from another. A gene-flow community is demarcated by barriers; a descent community is not (Chapter 7). While a descent community does have a real (genealogical) connection, the line between species is not defined by descent, but is set by artificial borderlines (Mallet, 1995).

Therefore, descent communities cannot be natural groups because they form a continuum, not a structured set of groups. Genealogical connection implies the birth of children or grandchildren, not the birth of taxa. Nothing in the succession of consecutive generations constitutes the end of one taxon and the beginning of another taxon (Simpson, 1961). The descent connection exists in nature, but there are no defined boundaries. The boundaries in the descent community must be created by human-imposed criteria. In contrast, gene-flow connections have natural boundaries and are therefore acceptable for use in forming a natural species concept.

6.5
The Concept of the Gene-Flow Community in Eukaryotes and in Bacteria

Lateral gene transfer is not restricted to eukaryotic sperm-egg fusion. Lateral gene flow also exists in prokaryotes (bacteria). The peculiarity of bacteria, however, is that the horizontal gene transfer does not only occur between related organisms, but bridges wide evolutionary divides. Bacteria that are evolutionary distant from each other can exchange genes (Dagan, rtzy-Randrup, and Martin, 2008). There is even occasional transfer between archaebacteria and eubacteria, which are assigned to different kingdoms because of their evolutionary distance (Woese, Kandler, and Wheelis, 1990).

Recently, an essay was published with the noteworthy title, "Species do not really mean anything in the bacterial world" (Hollrichter, 2007). This title signifies that there are no reliably covarying traits in bacteria that would allow a consistent scheme of classification. Bacteria can be classified according to one property, but another property may overlap across the units; classification by this property would result in other delimitations. By classifying according to one trait, one may, of course, obtain groups. However, by then classifying according to a second trait, one may obtain completely different groups.

After bacteria were classified by morphological criteria 150 years ago, the age of bacterial cultures began. Accordingly, bacteria were arranged by physiological and biochemical properties or by their "specific" pathogenicities. The "chemo-taxonomical" classification was based on fatty acid and sugar components. In the 1960s, however, a completely new classification was conducted according to genomic similarity. Carl Woese has given preference to the comparison of 16S rRNA gene sequences for creating taxonomic classifications.

As of today, about 7000 species of bacteria have been designated and classified into 1194 genera, 240 families, 88 orders, 41 classes and 26 phyla (Hollrichter, 2007). This classification is based on the following definition of species: two organisms belong to the same species if their genomes are more than 70% identical, their 16S rRNA sequences are at least 97% identical and their phenotypes are very similar. This definition, however, cannot be applied to most bacteria, as the three criteria are not consistent. Due to lateral gene transfer, there are bacteria whose genomes are more than 70% identical, which thus should belong to the same species, but which also exhibit large differences in their 16S rRNA sequences, and so should belong to different species. The reverse is also seen.

It follows that there are neither phylogenetic species nor typological species of bacteria because bacteria can be grouped only by individual traits and not by trait sets. It is not possible to classify bacteria by clusters of covarying traits (Chapter 3) because almost every individual trait suggests a different group formation.

Can bacteria be grouped into gene-flow communities, or are they “species-less” by this concept as well? The answer is no: there are gene-flow communities. Bacterial gene-flow communities are, to a certain extent, comparable with gene-flow communities in eukaryotes, because bacteria do not exchange their genes uniformly or arbitrarily. Instead, bacteria form groups that preferentially exchange their genes, although there are no distinct boundaries. Such groups can be called gene-flow communities. Such groups are delimited from other groups by having no or only a limited degree of mutual gene exchange with the other group (Dagan, rtzy-Randrup, and Martin, 2008).

It is possible, not only formally but also with respect to the population genetic and evolutionary consequences, to consider the bacteria that form a group that participates in mutual gene exchange as a kind of species that also can be called a gene-flow community; however, the processes of lateral gene transfer in eukaryotes and prokaryotes are biologically different. In eukaryotes, lateral gene transfer is sperm-egg fusion; in prokaryotes, it is conjugation (gene exchange via temporary cell fusion) or transduction (that is, gene exchange via phages). In both processes, gene transfer in prokaryotes differs from gene transfer in higher organisms in three remarkable biological aspects:

1) In prokaryotes, lateral gene transfer proceeds only in one direction (unilaterally). One of the two partner cells is the donor cell, and the other is the recipient cell. Essentially, this process is not a gene exchange at all, but a one-sided gene transfer. The donor cell does not receive any new genetic material and remains unchanged after conjugation, when the two partner cells separate. Only the recipient cell undergoes a genome alteration. This process is entirely different from the fusion of sperm and egg in eukaryotes, in which the two merging partners cannot be differentiated into a donor and recipient.
2) Gene transfer in bacteria is not a permanent fusion of two cells followed by a complete karyogamy, as in the sperm-egg merger of the sexual gene transfer in higher organisms. Bacterial conjugation always proceeds through a temporary cell fusion that disintegrates after the gene transfer. In transduction, phages enter the bacterial cell and introduce the DNA of another bacterium.
3) During bacterial lateral gene transfer, usually only a part of the genome is transferred from the donor into the recipient, rather than the entire genome of the donor cell. The fact that complete genomes are not transferred makes the bridging of large phylogenetic distances easier to understand.

6.6 Uniparental Propagation in Eukaryotes

The disadvantage of viewing the species as a gene-flow community is that, in nature, species such as these exist only in organisms with (lateral) sexual gene transfer. Only organisms that, in the course of their life cycle, transfer their genomic DNA into another organism are connected to each other in a group that we can call a species.

In contrast to biparental organisms, uniparental organisms are capable of only vertical gene transfer from parent to offspring. They are unable to transfer their DNA laterally from one phylogenetic lineage into another. They never exchange genes, and they never mix their genomes.

Uniparentality is the production of offspring from only one parent. There are three types of uniparental reproduction:

1) Vegetative reproduction, meaning the creation of daughter organisms from somatic mother cells (e.g., in hydrozoans or liver flukes).
2) Parthenogenesis, meaning the origin of a daughter organism from an unfertilized egg cell; that is, the development of a new individual from an egg without the addition of sperm (e.g., in rotifers, water fleas, plant lice or certain lizards). As parthenogenesis is a reproduction through germ cells, it is a sexual reproduction, although a uniparental sexual reproduction.
3) Self-fertilization, meaning the ability of certain hermaphrodites (which are individuals that produce both sperms and eggs) to inseminate their haploid eggs with their own sperm and so to successfully produce offspring without the contribution of a second parent. This happens often in plants (e.g., in barley, beans or peas), but rarely in animals (e.g., in tapeworms).

All uniparental organisms are species-less organisms with regard to the gene-flow community because they do not have any genetic connections with each other after birth. They do not form cohesive groups with mutual gene exchange. Uniparentally reproducing organisms only transfer their DNA to other organisms through the succession of consecutive generations (mother – daughter – grandchild). Thus, they lose any genetic connection to each other immediately after their birth. In uniparentally reproducing organisms, every birth would represent the origin of a new species in the sense of the gene-flow community. This notion, however, does not make any sense, as it is obviously expected of every species concept that it refers to the combination of several organisms in a group. Otherwise it would not be a species concept.

In some animal groups, such as hydrozoans or liver flukes, uniparental and biparental generations alternate during the life cycle. In other animal groups, such as rotifers and water fleas, uniparental and biparental reproduction alternate depending on ecological conditions such as temperature and humidity. In these cases of alternation of uniparental with biparental reproduction, the application of the gene-flow-community species concept is not questioned. Without a doubt, there is a lateral connection among the organisms within the species, even if the connection only appears at variously long intervals and sometimes only once every couple of years. Only those organisms with exclusive uniparental reproduction are species-less.

However, do such organisms exist? Although uniparental reproduction is common, for most such organisms, it is only a small part of the life cycle. Longer-term uniparentality seems to be rare in eukaryotes, at least in animals

(Ghiselin, 1997). Longer-term uniparentality is hard to substantiate, as an occasional biparental fertilization can only be recognized if the organisms are observed uninterruptedly for years. As such surveillance is generally impossible, bisexual stages may be overlooked, especially if they do not occur for years.

The bdelloid rotifers are famous for having persisted without biparental sex for millions of years (Welch and Meselson). There are no males. Only females are found, and these individuals are all diploid. As the homologous alleles of the diploid genome have diverged to an unusually high extent in DNA sequence, it is clear that genetic recombination between the homologous alleles has not occurred for many generations of evolutionary time. Under biparentality, the homologous alleles would have aligned their sequences through occasional recombination. As no exchange has occurred, it appears that there has been no genetic recombination for millions of years; however, even among the *Bdelloidea*, the rare occurrence of mictic stages cannot be ruled out with complete certainty (Ghiselin, 1997).

The deep skepticism regarding permanent uniparentality in an animal species results from the awareness that the loss of bisexuality in particular phylogenetic lines should apparently come to a dead end in the evolutionary long run. Apparently, sexuality cannot be reinvented repeatedly in nature. There are no indications that a *de novo* origin of sexuality has ever occurred in phylogenetic lines that have lost sexual reproduction (Ghiselin, 1997).

Biparental sexuality apparently is an ancient biological process that originated in the common ancestors of all currently living eukaryotes and has since been maintained.

The conclusion that organisms with permanent uniparental reproduction are species-less because there is no interorganismic connection is unsatisfying. A taxonomist wants to classify the entire diversity of life and not only a subset of organisms, but nature is not made to be classified by humans (Sterelny and Griffiths, 1999). Only the need for order forces us to find a place for all and everything.

One interesting example in this context is the species status of "workers" in many social *Hymenoptera* (bees, wasps, ants). These workers are sterile and so do not have any reproductive connection to each other; however, it makes little sense to consider the sterility of *Hymenoptera* workers, which are dead ends of gene flow, as a species problem. This conclusion results from the following reasoning: one can imagine a beehive as a "super organism," in which the reproducing queen creates workers that are doomed to die off, like somatic cells split off from the germ line (Tautz, 2008). Like worker bees, somatic cells are end lines of development no longer capable of reproduction. Only the germ cells continue living. Similarly, the sterile *Hymenoptera* workers die, but the queens are inseminated and continue to reproduce.

The distinction between an ontogenetic genealogy of cells and a genealogy of organisms should not be as strong as it currently is. In animals and plants, there are many transitions from multicellular organisms to colonies of organisms that are connected like the cells in an individual organism. Examples for these "super organisms" are the animal colonies that are formed by many hydrozoans and corals. The next-higher level would then be the beehive.

6.7
Why do the Individuals of a Species Resemble Each Other?

Why do the organisms of a species resemble each other? There are three possible ways to explain this.

1) First, resemblance can be the result of a common descent. Children resemble their parents because of descent; however, mutation and selection cause increasing alteration of traits. As a consequence, genetic and phenotypic heterogeneity grow over time. Finally, over long evolutionary stretches, the more or less homogeneous appearance of the individuals of a species can no longer be explained by their common descent.
2) Secondly, the homogeneous appearance of the organisms of a species may result from a repeated stabilizing of the genomes via sexual genome fusion. Thus, continuing biparental sexual contacts between organisms prevent genomic divergence. If a group of organisms at the periphery of the distribution area of the species were to adapt to local environmental conditions, these divergent developments would be reversed by backcrossing with the organisms living at the center of the distribution area. How strong are these forces? How often and across which distances are the genomes of distant organisms stabilized by sperm-egg fusion and genetic recombination? Can newly mutated alleles really reach all distant areas of the species range? Does the well-known phenomenon of allele fixation really reach all distant populations of a species?

   The cofounders of the gene-flow-community species concept, Theodosius Dobzhansky and Ernst Mayr, put substantial weight on the homogenizing force of genetic recombination. In sexual gene flow, they saw a main reason why species actually exist (see the Section "The 'why' of speciation" in Mayr, 2000). The 1930s and 1940s were the time of the "Great Synthesis," which was a joining of evolution, genetics and taxonomy. Gene flow among the organisms of a gene-flow community was not only perceived as a connecting force that keeps the individuals of a species together, but also as the cause of the organisms' identity of traits across the entire range of the species. Gene flow was thought to cause the organisms in a community to have an overall homogeneous appearance and thus belong to the same species (Dobzhansky, 1937; Mayr, 1942). The morphological and physiological integrity of a species was thought to be created and maintained by gene flow.

   The reasoning behind this was as follows: because of local adaptations in different regions, the subpopulations in the long run would follow separate evolutionary pathways and thus would become increasingly divergent. Backcrossing, however, would occur repeatedly. Consequently, genomic interchange among the distant organisms would result in homogeneity of the genotypes. Because of genetic recombination, in principle, all allelic variants would reach all of the genomes in a population of a species. If some organisms were to diverge by mutation and selection, these deviations would again be reversed. Biparental sexuality would hinder this divergence among the organisms of a gene-flow community, maintaining homogeneity. Sexual reproduction was thought to be responsible for the preservation of traits that is observed in species in nature.

Such considerations rest on the assumption, however, that gene flow is "strong" throughout the entire geographical range of a species. By "strong" gene flow, it is meant that the alleles of an organism can reach geographically distant organisms within reasonable time intervals to prevent divergent evolution. The independent evolution of distant organisms can be prevented only if this assumption is true. Adaptations to local environmental conditions would be limited because the genomes would always be reintegrated into a homogeneous pool by backcrossing; however, this assumption is rooted in the purely theoretical speculations of the fathers of the "Great Synthesis." It was not experimentally based and could not have been by the standards of the day.

Even today, there is hardly any data on the strength of gene exchange between geographically distant populations of widespread species. There are considerable knowledge gaps concerning the actual homogenizing "force" of gene flow between distant populations of a species. It appears that the extent of internal genetic connection among the individuals of a biparental species has been largely overestimated in the past (Mishler and Donoghue, 1994). For example, the Willow Tit (*Parus montanus*) has spread continuously across the entire Eurasian continent, from Western Europe to East Asia. Presumably, the individual populations are connected to each other, without interruption, from France to the Pacific Coast (del Hoyo, Elliott, and Christie, 2007). Assume that an individual allele mutates in a Willow Tit in France. It would be interesting to know how many generations it takes this allele to travel via sexual contacts from Paris to Wladiwostok, but there are no data on this subject.

Ehrlich and Raven had already discovered in 1969 that the geographical range of many animals stretches across only a few kilometers and that the spread of the pollen and seeds of some plants is even restricted to only a few meters (Ehrlich and Raven, 1969). Examples include the common grass species *Festuca rubra* and *Agrostis tenuis*. Both species are composed of populations between which gene flow is extremely restricted. The pollen is apparently never carried across great distances, allowing for the establishment of local adaptations to minihabitats.

However, if individual alleles do not spread across large geographical expanses in several plant or animal species, then gene flow can only have a small or no influence on the preservation of the phenotypical integrity of the organisms of these species (Ereshefsky, 1999). Hence, what prevents the divergence of the individual populations of a species? Why do the organisms of a species look so similar across great distances?

3) There is a third possible explanation for the homogeneous appearance of the organisms of a species: selection pressure. Selection pressure may be the main force keeping the organisms of a species so similar. If a species has emerged, its traits are optimally adapted to its environment at that time. Deviations from this balanced pattern of adaptive traits would be eliminated by natural selection. This process seems to be the main reason why an organism cannot continuously change over time and across large geographical distances (Ereshefsky, 1999;

Dupre, 1999). All traits of an organism must function in a concerted fashion. This view considers the species as a balanced unit that is resistant to change. Selective pressure could thus be much more crucial than gene flow (Mallet, 2006).

That selection could be a major factor in maintaining the homogeneity of a species along large geographical distances already follows from the fact that uniparental organisms also produce distinct groups of phenotypically similar individuals. If biparental gene flow were the main factor underlying the homogeneity of a species, then uniparental organisms could not generate large populations of similar individuals in the long term. Therefore, the constancy in phenotypic appearance of the individuals of a species must not result mainly from gene flow (Lande, 1980; Mallet, 2006).
Perhaps this phenomenon is comparable to vehicles, which also remain more or less similarly shaped in the longer term because they cannot freely vary in form. Trucks, cars and motorcycles have many traits that must be aligned, as they are needed to form a functioning whole (Barton, 1993), and this integrity of form is independent of whether such a vehicle is required to function in the north or the south of Europe. In roadless terrain, however, other vehicle shapes prove superior, and four-wheel-drive, off-road vehicles and tracked vehicles are predominant.
However, the assumption that selection is the main factor that maintains species homogeneity should not be overstressed. The divergent adaptations of organisms to local environments are evident. Many traits are not homogeneously distributed across the entire range of a species. If this were not the case, there would be no races (Chapter 5).

6.8 Isolation by Distance

When alleles do not reach geographically distant organisms and thus evolve independently, this phenomenon is called "isolation by distance." Isolation by distance must be clearly distinguished from allopatry (see below). Two populations are considered allopatric if they are completely separated by external, usually geographical barriers, so that no gene flow is possible between them. If, however, two populations were geographically distant but still clinally connected by intermediary populations, then this situation would qualify as isolation by distance.

Isolation by distance is problematic because the organisms of distant populations of a species can lose their ability to crossbreed (Sterelny and Griffiths, 1999). If race *A* overlaps in range with race *B*, and race *B* overlaps with race *C*, then race *C* will not necessarily be cross-fertile with race *A* (Figure 6.2a). The ability of individuals in distant populations to crossbreed may become lost despite incremental clinal transitions along a chain of populations.

An example of the decreasing ability of distant organisms of a species to crossbreed was found in the 1940s and early 1950s by the British evolutionist and lepidopterologist E. B. Ford (Ford, 1940; 1954). Ford showed by the example of the Satyrid butterfly *Coenonympha tullia* (Large Heath) that with an increasing geographical distance between populations, the number of fertile offspring from crossings decreases. This early discovery should actually have had a greater influence than it did on the understanding of the species as a reproductive community in the sense of Ernst Mayr because it strongly suggests that not all individuals of a species are able to successfully crossbreed with each other. Ford's findings, however, like the discoveries by Ehrlich and Raven on the effective reach of pollen and seeds in plants (Ehrlich and Raven, 1969), have not received much consideration subsequently, as Mayr's opinion was predominant.

The suggestion that geographically distant populations of a species may lose their genetic compatibility is clearly demonstrated by the case of the "ring species" (Figure 6.2b) (see below). Ring species are continuously connected chains of races of a species that surround a geographic region than cannot be occupied because it is a mountain massif, an ocean or the polar region. Therefore, they show a ring-shaped distribution around the inhospitable region. At the endpoints of this geographic ring, the most distant populations encounter each other secondarily. Here, they are no longer genetically compatible and are unable to crossbreed. Although they still belong to the same species, they have become genetically different because their common ancestor dates back far into the past and the mutual exchange of alleles has become scarce or has ceased entirely.

Ring species are usually considered to be exceptions because geographically distant populations rarely encounter each other again under natural conditions at another geographic location. The ring species, however, is only a special case of a much more general phenomenon. That general phenomenon is isolation by distance (Figure 6.2a); the idea being expressed here is that the distant populations often are no longer compatible with each other genetically, even though a connection through incremental gene flow is still present. Distant organisms are often no longer able to crossbreed when the geographical range of the species is large. "Ability to crossbreed" means a long-term, stable ability to crossbreed over many generations, with the descendants competing with purebred descendants. In this sense, the ring species is not an exceptional form of the species, "superspecies" or even the manifestation of an incipient speciation. The ring species describes only a peculiar circumstance in which the distant populations encounter each other in nature. Apart from this peculiarity, ring species do not differ from most other widely spread species.

6.9 A Decrease in Lateral Sexual Gene Flow, together with Local Adaptation, Creates Races

Local geographical populations that diverge with regard to their traits and that therefore are clearly diagnostically different are called "races" or (synonymously) "subspecies" (Chapter 5). Races originate from different adaptations to the environmental conditions at different geographical locations. Races develop in the course of evolution, emerging when the "homogenizing forces" of lateral sexual gene flow become weaker than the force of selection driving adaptation to local are genetically based, the populations living in northern latitudes differ in genotype from the more southerly breeding birds, although the populations are clinally connected through overlapping zones and thus belong to the same species. The question is whether the populations can be crossed under these conditions.

The same traits hold for bird species whose populations breed at higher mountain altitudes and at the lower levels in the valleys. The hormonal control of nest-building, egg maturation and egg deposition must be regulated differently in the populations breeding in higher altitudes than in the birds that breed at lower levels, although the locations do not differ in the day-night ratio, the external trigger for egg laying and breeding activity. For Blue Tits (*Parus caerulescens*) on the island of Corsica, birds in the summer-green Downy Oak (*Quercus pubescens*) forests of the higher mountain altitudes breed four weeks later than their conspecifics at lower elevations, in the evergreen Holm Oak (*Quercus ilex*) belt (Blondel *et al.*, 1999).

The genetic foundation of such different local adaptations has repeatedly been substantiated. Organisms from populations with different periods of activity during the year retain this rhythm after translocation to other locations (Helm, 2009); they do not adapt to the new environmental conditions. This stability has not been shown in birds only; it also applies to other animals. In Germany, specimens of the flightless Ground Beetle *Carabus auronitens*, for example, have been translocated from the forests of Münster/North Rhine-Westphalia to the forests near Arnsberg (only 80 km away) (Schwöppe, Kreuels, and Weber, 1998). The beetles occur in both places, but the populations have different seasonal periods of activity. After the translocation, the beetles retain the annual rhythm of their region of origin and do not survive long.

In the course of experiments like these, the question arises as to whether the geographically different populations can successfully be crossed with each other, so that their offspring survives under natural environmental conditions. There are not many investigations of this subject. As always, the question is not whether such crosses can be accomplished experimentally in the laboratory, but whether they would occur under the conditions of natural mating behavior and, most of all, whether the resulting offspring would be sufficiently fit and fertile to compete with the offspring of pure-bred individuals. As the adaptations have a genetic foundation and many different traits are involved, crosses among the geographically different populations should recombine the genes for local adaptations by genetic crossing-over, thus disrupting genes that must remain coupled to function properly. If they are recombined by cross-breeding, the mixed offspring should possess a mixture of the adaptive traits. Such mixed types should have little chance of competing in the long term against the incumbent types. Incorrect, intermediate times of egg deposition and breeding would likely manifest.

6.11
Are Migratory and Sedentary Birds Able to Crossbreed?

Many bird species consist of populations of both migratory and sedentary birds (Chapter 5). This is very often then the case if the birds have a large geographical range and breed in the north of Europe or Asia, as well as the south. In the south and also in the west, for example, in the Mediterranean region and in Ireland and Great Britain, a sufficient food supply is often available throughout the entire year. For this reason, the birds can stay in the breeding habitat in winter and do not need to give up their territories, remaining sedentary. Populations of the same species, however, also breed in the north and in the east. Here, food shortages in winter would cause the animals to starve if they did not leave their breeding grounds during the winter months. These are migratory birds.

Most bird species, however, do not leave their breeding grounds when food becomes scarcer, but depart significantly earlier. Most migratory birds leave Central Europe in July, August or early September, when food is still abundant. Thus, they reach their distant winter quarters in due time. From this observation, it follows that the migratory behavior is not controlled environmentally by hunger or cold, but by genetic factors that respond to day length, for example. The birds are not free to decide whether they will migrate in fall or stay at home; instead, they are controlled by their genes. At least in the case of long-distance migratory songbirds that often are not guided by their parents during migration, all traits that distinguish migratory from sedentary birds must be genetically controlled.

The differences between migratory and sedentary birds are enormous. As in the other cases in which many traits distinguish different populations of a species, the question arises as to whether these traits are controlled by just as many genes (Helm, 2009). Regarding the many differences between migratory and sedentary birds, it is difficult to understand how these organisms would be able to crossbreed. If there are many genes involved, it can hardly be imagined that the sedentary populations of a bird species would be able to crossbreed with the migratory populations of the same species because the genetic incompatibilities of the different alleles should be too large. The alleles for migratory or sedentary traits would recombine in the hybrids and lose their genetic linkage. The offspring of such crosses would show a mixture of traits belonging to migratory and sedentary birds.

Indeed, experimental crossings have shown that hybrids from migratory and sedentary birds exhibit properties that are intermediate in regard to the different traits (Pulido *et al.*, 2001; Helm, 2009). The many crosses between migratory and sedentary birds show that at least some traits that distinguish migratory from sedentary birds were inherited in an intermediate manner (Berthold and Querner, 1981; Helm, 2009). The hybrids resulting from such crosses have some alleles for migratory behavior as well as some alleles for sedentary behavior. They obviously are hampered in fertility and fitness.

The crosses between migratory and sedentary birds have, however, been conducted under captive conditions. The generation of phenotypic intermediates appears to contradict the realities seen in nature, where environmental influences interfere with genetic constitutions (Helm, Fiedler, and Callion, 2006). In nature, the clinal transitions between migratory and sedentary birds appear to contain few intermediates. Instead, birds that leave the breeding grounds in fall live alongside birds that stay behind during winter.

In nature, several migratory and sedentary birds apparently crossbreed without a problem in the clinal intermediate populations of partially migratory birds. How can this be? The question is whether, under natural conditions, the hybrids of migratory and sedentary birds would be able to adapt to the environmental constraints on changing their inherited migration directions and distances (Helm, Fiedler, and Callion, 2006).

6.12
Are Geographically Distant Populations of Stonechats (*Saxicola torquata*) or Blackcaps (*Sylvia atricapilla*) Genetically Compatible?

Well-investigated examples of genetic differences between distant populations include Stonechats (*Saxicola torquata*) and Blackcaps (*Sylvia atricapilla*).

Stonechats have a large breeding area across Eurasia and Africa. Central European Stonechats are short-distance migrants that winter in the Mediterranean region. They return to their breeding grounds in March or April and produce two or three broods before they depart in October. On the British Isles, the birds start laying eggs relatively early and may initiate three clutches over a long breeding season. Siberian Stonechats from Kazakhstan breed at latitudes similar to those in Europe; due to the continental climate, however, their reproductive period is short. After long-distance migration from their South Asian winter quarters, Siberian Stonechats arrive at the breeding grounds in May. They usually raise only a single brood and leave their territories soon after breeding. Hence, Siberian long-distance migrants spend only half as much time on the breeding grounds as do European short-distance migrants. African Stonechats in equatorial Kenya are residents. The birds maintain year-round pair territories and usually lay a single clutch at the beginning of the first rainy season (Helm, 2009).

The reproductive timing of the different populations is genetically fixed and controlled by circannual rhythms and photoperiodism. The problem, however, is that in the different geographical regions, very different day-night parameters induce breeding. Birds from four different regions (Central Europe, Ireland, Siberia and Kenya) were transferred to Germany and reared under the same circannual conditions in captivity (Helm, 2009). As expected, they did not adapt to the new conditions, but retained their original population-specific breeding windows.

The crossbreeding of birds with different schedules resulted in a distinct loss of reproductive success because the breeding partners were unable to adjust their schedules to those of their mates. Thus, geographically distant individuals have lost a large part of their mutual reproductive compatibility (Helm, 2009).

The situation with the Black-capped Warblers (*Sylvia atricapilla*) is similar. Blackcaps are also widespread on the palearctic Northern hemisphere, breeding from the Cape Verde and Canary Islands through Spain, Central and Northern Europe and Siberia. On the Cape Verde and Canary Islands, they are sedentary birds; in the Mediterranean region, they are partially migratory birds, and in the other regions, they are purely migratory birds.

The Blackcaps might consist of several genetically incompatible populations throughout their geographic range. This possibility becomes evident among populations that live geographically much closer. The Blackcaps breeding in the region of Lake Constance in southwest Germany and those breeding around Lake Neusiedl on the Austrian-Hungarian border live only about 600 km apart from each other. Nevertheless, the two populations migrate in completely different directions in late summer. The southwestern German Blackcaps migrate to the southwest in late summer and stay in their winter habitat in Spain. Then, their migratory restlessness subsides. The southeastern Austrian Blackcaps migrate from their breeding grounds to the southeast and require several weeks to arrive in their winter habitat in East Africa. Only then does their migratory restlessness expire.

Neither the correct migratory direction nor the information on arrival in the final winter habitat can be learned from parents or from conspecific companions, as Blackcaps migrate alone and at night. Instead, this information must be genetically founded. The significant difference in this genetic information between geographically distant populations raises the question of how crossings between distant populations can be possible, as genetic recombination should cause the resulting allelic combinations in the F1-hybrid to be incompatible.

Indeed, large numbers of experimental crossings between Eastern European and Western European Blackcaps have shown that mixed properties result (Berthold and Querner, 1981). The hybrids of the Eastern European and Western European Blackcaps display in part the migratory properties of the southeast migrants and in part those of the southwest migrants. Both migratory directions and migratory durations seem to be inherited in an intermediary fashion in the hybrid (Pulido *et al.*, 2001); however, all of these crossing experiments have been carried out in captivity. It is not known how the hybrids would behave under natural conditions. The behavior of hybrids like these has not been tracked in the wild.

The problem is that the Western European Blackcaps are linked uninterruptedly to the Eastern European ones. Thus, the question of transition emerges. Individuals living in the region of transition are very well able to crossbreed. It is not easy to explain how the behavior of those organisms is genetically controlled. It cannot be that the migratory direction gradually changes from southwest to southeast in the organisms of the clinal transitory region; the solution must be binary.

The open question of successful crossbreeding of the different populations presents a problem for the species status of Blackcaps. If birds with different genetic drivers of migratory directions or migratory durations were genetically incompatible, this difference would be a clear postzygotic mating barrier (see below). In the sense of Mayr's species concept, the different Blackcap populations would then be different species. Following the gene-flow-community species concept as defended in this book, however, genetic incompatibility between distant populations would not be a problem for the species status, as long as the gene flow is not interrupted.

In the recent literature, the first indications of the existence of prezygotic mating barriers were published (Rolshausen, Hobson, and Schaefer, 2010). Within the last 50 years, Blackcaps from southwest German breeding grounds have evolved a novel migratory strategy. An increasing proportion of these birds now winter in Britain, rather than migrating to the traditional sites in the Mediterranean area. The individuals that now winter in Britain migrate to the northwest. They still live in overlapping regions with the individuals that migrate to the southwest and winter in the Mediterranean region. The first signs of isolating barriers between the two populations are observed; these barriers may lead to the evolution of assortative mating (see below). Such behavior would indeed constitute the early stages of speciation.

6.13
Are Univoltine and Bivoltine Butterflies Able to Crossbreed?

Roughly comparable with the intraspecific dimorphism of migratory and sedentary birds are univoltine and bivoltine butterflies. Univoltines are butterflies that produce only one imaginal generation per year. Bivoltines generate two successive generations within a year. In most cases, bivoltine butterflies hibernate as pupae, the imagines hatch in spring, and their offspring produce a second imaginal generation in summer.

As with migratory and sedentary birds, univoltine and bivoltine butterflies are adapted to different latitudes. In Southern Europe, many butterflies produce two or even three generations each year and therefore occur in large numbers in this region. In Northern Europe, however, the butterflies of this same species have only one generation and are accordingly more rare (Rensch, 1951).

For example, the Swallowtail (*Papilio machaon*) (Color Plate 1) is univoltine in Northern Europe. In the course of the year, only a single generation of butterflies (imagines) hatches, and the offspring then hibernate as pupae. In Central Europe, the Swallowtail is usually bivoltine. The first generation's imagines fly in May; their offspring pupate in June, and the pupae still hatch in the same year in July. These butterflies represent the second generation. This generation lays eggs in July, and the pupae developing from these eggs hibernate from September to late April; however, in an unusually warm summer, a third generation is observed in Central Europe. Still farther to the south of Europe, the Swallowtail regularly produces three generations per year.

The Swallowtail can only undergo the winter diapause in the pupal stage. Eggs, caterpillars and imagines are not resistant to longer periods of frost. Thus, no other developmental stage can hibernate. Only the pupa can do so.

From this phenomenon, it must be postulated that uni-, bi- and trivoltinism must be founded on genetic differences. If these differences were induced by the temperature or other external factors, in an extreme case, the existence of entire species could be endangered because the wrong developmental stage could be obliged to hibernate. Eggs or caterpillars, however, cannot survive the cold season.

Assume that in an unusually warm summer, the Swallowtail pupae of the bivoltine population in Central Europe might "decide" to hatch again in late August and produce a third generation, rather than wait for the winter. If this were to happen, the caterpillars of this third generation would live in dangerous conditions. If the temperature were to drop immediately in September, the caterpillars would not reach the pupal stage. As the caterpillars cannot hibernate, they would perish. All Swallowtails in Central Europe would become extinct if they all were to react to a warm August in that way.

Therefore, one must conclude that these butterflies do not all react to a warm late summer in the same way. Instead, their ability to produce a third generation cannot be triggered by environmental conditions alone. Instead, the trivoltine Swallowtails in Central Europe must be a genetically different population in comparison to the bivoltine Swallowtails. Only the members of the trivoltine population hatch three times a year. The members of the bivoltine population are genetically programmed not to hatch in late August, even if it is warm (as in a Mediterranean environment). This conclusion would explain why the species survives in Central Europe.

Another example is the Common Blue Butterfly (*Polyommatus icarus*) in Sweden (Color Plate 1). This species consists of a univoltine population in the north and a bivoltine population in the south of Sweden. Nygren, Bergström, and Nylin (2008) have shown that these two populations are genetically different in terms of a fairly large number of traits. It is not possible to explain uni- or bivoltism as a reaction of individuals to external climatic conditions.

The butterflies in the north have different metabolisms and significantly slower development. They need more time to finish development of the egg, the caterpillar and the pupal stage, as they must produce only a single generation. The butterflies in the south develop faster because they must manage two generations per year. Common Blues transferred from the north of Sweden to the south did not change their behavior in reaction to the warmer climate. Instead, they maintained their slow development and produced only one generation per year. The day-night rhythm and other climatic conditions of the south, such as temperature, did not cause the northern populations to transition into bivoltinism (Nygren, Bergström, and Nylin, 2008).

These experiments clearly show that bivoltism and univoltism are genetically driven. Several genetic differences characterize and differentiate the Common Blue into two different populations. As with migratory and sedentary birds, this finding again raises the question of whether univoltine and bivoltine populations of butterflies would be genetically compatible if they were crossed. As clinal transition regions exist in which univoltine and bivoltine individuals coexist, as with partially migratory birds, it is tempting to assume that univoltine and bivoltine butterflies are different morphs of the same species (Chapter 5). The frequency distribution of the morphs differs between different geographic regions. In the northern populations, almost all of the alleles correspond to univolism; in the southern populations, almost all of the alleles correspond to bivolism, and in the overlapping geographic region, there are populations where both morphs live side by side.

6.14
Speciation Genes, Pre- and Postzygotic Barriers

Reproductive incompatibility is the result of the mutual incompatibility of the alleles of particular genes (Wu, Johnson, and Palopoli, 1998). Reproductive incompatibility is not a straightforward consequence of evolutionary divergence, which is the spatial and temporal distance of the populations from each other. No biological law necessitates that phylogenetically distant populations must "automatically" be reproductively incompatible. The organisms in evolutionary distant strains become reproductively incompatible only by chance, not by any biological law, as the largest proportion of the genes has nothing to do with species barriers.

It would be too easy to interpret the situation such that, in the case of reproductive incompatibility, the two populations just had diverged "far enough" from each other that they were no longer genetically compatible. Genetic incompatibility is not a necessary consequence of the divergence between two populations. Instead, reproductive incompatibility is the result of the mutual incompatibility of the alleles of a few particular genes, the interactions of which lead to restrictions on vitality or fertility. Such genes are called "speciation genes."

Speciation genes are responsible for speciation. They are responsible for keeping two species segregated from each other. These genes cause the splitting of species. They include, for example, the genes that control the mating choice. Their task is to prevent mating with different species. Speciation genes also control adaptation to species-specific food plants, habitats or climates. It is an important task of modern taxonomy to find genes of this kind (Wu, Johnson, and Palopoli, 1998).

The statement that speciation is primarily founded on very few particular genes, not on overall genetic divergence, is substantiated by the example of the New World Ruddy Duck (*Oxyura jamaicensis*) and the Old World Ruddy Duck (*Oxyura leucocephala*) (see below). The species were separated from each other for a long time. They are phylogenetically distant and became phenotypically very different; however, after the New World Ruddy Duck was introduced to Europe, it began to hybridize with the endemic European Old World Ruddy Duck. Presently, the latter is seriously endangered as a separate species because there appears to be no hybridization barrier. Despite a long period of separation, no genetic incompatibility has evolved between these two species. If the further expansion of the New World Ruddy Duck cannot be stopped, the Old World Ruddy Duck may become extinct by hybridization in the next few decades.

Speciation genes must be separated into prezygotic and postzygotic speciation genes. Prezygotic speciation genes are those that control the processes that are responsible for the final goal of sperm-egg fusion and, therefore, zygote formation between two organisms. For this reason, they are called prezygotic genes.

In the case of external fertilization, prezygotic speciation genes code for signaling molecules that allow sperm cells to fuse with the eggs of the correct species or that prevent sperm cells from fusing with the eggs of the wrong species. Furthermore, they code for receptor molecules on the egg's surface that control the fusion of sperm cells with egg cells of the correct species. These receptor molecules could also inhibit zygote formation between different species.

In the case of internal fertilization, prezygotic speciation genes control copulation. If two organisms belong to different species, the prezygotic speciation genes prevent a successful copulation. Therefore, in the case of internal fertilization, the term "prezygotic barrier" is almost synonymous with the term "premating barrier." For example, prezygotic speciation genes regulate species-specific plumage and song traits in birds, which serve partner identification and stimulate the willingness of the female to mate. Prezygotic speciation genes cause assortative mating; assortative mating refers to the selective choice of a sexual partner of the correct species and a sexual aversion towards the members of a different species. Prezygotic barriers also include structural traits of the copulation apparatus; for example, in the case of many insects, there are purely mechanical reasons to not be able to copulate with a foreign species. In many cases, prezygotic barriers are a complex combination of several distinguished behavior patterns that interact between the mating partners. That is the reason why prezygotic barriers often cannot be reproduced under experimental conditions with animals in captivity.

In addition to prezygotic barriers, there exist postzygotic barriers that also prevent crossing among the members of different species. Similar to the prezygotic barriers, the postzygotic barriers are regulated by specific genes, called postzygotic speciation genes. Postzygotic barriers are those that become noticeable only after hybrid formation has already occurred. In other words, postzygotic barriers can only be effective if prezygotic barriers do not exist or if the prezygotic barriers were imperfect ("leaky"). Postzygotic speciation genes are responsible for any type of hybrid dysgenesis, which means properties of the F1 offspring that disadvantage a hybrid, compared to purebred offspring.

In many cases, species hybrids are not able to effectively compete with purebred offspring with respect to their vitality and/or fertility. In extreme cases, a reduction in vitality appears already in early embryonic development; the coordinated development of specific cells or tissues is deranged. In other cases, however, postzygotic barriers become noticeable only after several generations because the reduction in the vitality or fertility is faint and is not easy to observe. The result only has an impact from competition with purebred rivals. That is the reason why postzygotic barriers often cannot be reproduced under experimental conditions with animals in captivity.

6.15 Hybrid Incompatibility

Hybrid incompatibility is one of the possible reasons why two organisms belong to different species. However, it has to be considered that that are many examples that isolation by distance also includes hybrid incompatibility. Alternatively, it has to be considered that that are many examples of allopatrically separated organisms that are not incompatible. Hence, hybrid incompatibility is not the same as speciation.

Until today, little has been known about genes that lead to allelic incompatibility in a species hybrid and, thus, about postzygotic disturbances. Little is known with respect to how many genes are involved and what, in each case, leads to the discrepancies. It is not very scientific to say that phylogenetically distant organisms have simply drifted far apart from each other genetically and that the drift causes the hybrids to be reduced in vitality and fertility. Instead, it should be one of the first tasks of the taxonomist to find out why the genomes do not cooperate anymore because that lack of cooperation is one of the possible reasons that two individuals belong to two different species.

A few genes have already been identified that are responsible for hybrid incompatibilities (Coyne and Orr, 2004). The difficulty of finding genes that are responsible for hybrid incompatibility is of a fundamental nature. This difficulty arises from the fact that, for example, two related *Drosophila* species would have to be crossed with each other to find the genetic causes of their cross-incompatibility. For a moment, this action sounds like a contradiction in itself (Wu, 1996).

More recent investigations on *Drosophila* and other organisms indicate that, in the case of hybrid incompatibilities, the fertility of the species hybrid is more strongly affected than its vitality. In most cases, reductions in fertility belong to the restrictions that first become noticeable in the hybrid, while the hybrids' vitality is not yet compromised. Only after even more incompatible crossings, vitality loss appears in the offspring as a further reduction in fitness. This remarkable difference in the appearance of fertility versus vitality reduction is apparently founded in the fact that fertility is regulated by genes that evolve at a faster rate than the genes that control vitality (see Haldane's Rule, below) (Coyne, Simeonidis, and Rooney, 1998; Wu *et al.*, 1995).

Yet one point of view is certain: the problem of hybrid incompatibility is a consequence of very specific gene expressions. It is very imprecise to state that phylogenetically distant organisms have simply drifted far apart from each other genetically and no longer match only for this reason. Even genomes that are phylogenetically distant to a relatively far extent can, in some cases, build a hybrid organism that is still vital and fertile (Turelli, Barton, and Coyne, 2001). Conversely, genomes that are phylogenetically related more closely can, in many cases, exhibit nonviable incompatibilities.

Even within the same gene-flow community, and thus, in closely related organisms, there could be certain genetic incompatibilities. This scenario already appears via the numerous natural abortions within an otherwise functioning gene-flow community. In the case of humans, only a portion of the fertilized egg cells develop into a viable child (Grobstein, 1979). These data make clear that, in a gene-flow community, not all of the sperms and eggs, and thus also not all of the organisms, are reproductively compatible with each other, either factually or potentially. The frequent occurrence of reproductive incompatibilities within a species again supports the view that the species, if defined as a gene-flow community, is much more precise than the notion that the species would be a reproductive community. Many members of a species are not reproductively compatible with each other. Thus, it is not possible to test for species membership of organisms by crossing selected individuals.

Postzygotic incompatibilities are, therefore, not a matter of individual, so-called statistically selected organisms. They are a property that occurs in populations and affects the majority of the organisms present, but cannot, in each case, be applied to the single individual. Even the horse and donkey, which are textbook examples of postzygotic incompatibility, can, in individual cases, be fecundly crossed with each other. This already follows from the fact that there are domestic horse races, into whose bloodline donkeys have been bred. It would not be consistent to prove the species status of a population by choosing single individuals by chance, testing them for their fertility and then inferring the species status from the result (see the problem with allopatry below).

Reproductive incompatibility is different from phylogenetic distance. Thus, the definition of a species as a group of organisms that is separated from another species by a specific phylogenetic distance is not the same as the definition of a species as a group of organisms that is reproductively isolated from another species (see the criticism on barcoding in Chapter 4).

6.16 Haldane's Rule and the Genes for Postzygotic Incompatibility

More than 80 years ago, the British geneticist John Burdone Haldane discovered a remarkable fact. If two species are crossed with each other, then postzygotic deficits in the hybrids of the F1-generation are expressed much more strongly in the heterogametic sex than in the homogametic one. The heterogametic sex is the sex that, instead of two X chromosomes, possesses the X/Y configuration or the X/0 configuration. For this reason, the heterogametic sex produces two types of mature germ cells during meiotic division: one half are X gametes and the other are Y or 0 gametes. For this reason, it is called the heterogametic sex. These individuals are usually the males. However, in some animal groups (well-known examples are the birds and the butterflies), it is the female that is the heterogametic sex. In the latter case, the sexual chromosomes are given different designations: they are not called X and Y chromosomes but are instead called Z and W chromosomes. The introduction of these additional terms is irritating. The designations Z and W instead of X and Y are superfluous and increase confusion because the sexual chromosomes have the same meaning genetically, whether they occur in the male or in the female.

Haldane discovered that in species crossings, the X/Y sons more frequently exhibit damage than the homogametic X/X daughters. This discovery entered biology as "Haldane's Rule." It was then that the American geneticist Hermann Joseph Muller focused on this rule in the 1940s and offered a plausible and simple explanation for it. Because the heterogametic offspring inherits a complete set of autosomes plus an X chromosome from the one parent, but it inherits from the other parent no sex chromosome (in the case of the X/0 type) or only the Y chromosome, which contains few genes (in the case of the X/Y type), this scenario would lead to an imbalance between the autosomal genes and the sex-chromosomal genes in the X/Y or X/0 offspring. The autosomes stem half and half from both parent species; the sex chromosomes stem completely or almost completely from only one parent species. This imbalance was called X-autosomal disharmony of the heterogametic hybrid and was used for half a century as an explanation of Haldane's Rule. X-autosomal disharmony was deemed to be a plausible interpretation with regard to the postzygotic deficiencies in the heterogametic offspring of species crossings. Once again, something troubling scientifically becomes apparent in the history of biology: Muller's hypothesis of X-autosomal disharmony was convincing only because of its simplicity and plausibility and not because of experimental support for it.

Only decades later, the general validity of Haldane's Rule was disproved by new observations and experiments. The observations and experiments were conducted with simple methods. Initially, it was substantiated that the entire matter is far more complex. First, Haldane's Rule affects fertility to a greater extent than vitality. The hybrid offspring is much more likely to be sterile than to have its vitality be weakened. Second, there is a distinct difference in whether the heterogametic hybrids are male or female. If the sons are heterogametic (X/Y constitutions), then the decrease in fertility is affected more significantly than if the daughters are heterogametic (Z/W constitution) (Wu, Johnson, and Palopoli, 1998). This difference between the sexes could not be explained by a simple understanding of Muller's explanation.

However, the universal validity of Haldane's Rule was even more powerfully disproved by a simple experiment on *Drosophila*. By a genetic trick, it was possible to produce female F1-species-hybrids that received both X chromosomes from only one parent species. Thus, these experimentally produced homogametic daughters have the same X-autosomal imbalance as the hetrogametic F1-sons of hybrid crosses. Despite this scenario, the daughters were, in part, fertile, while the male offspring, having a similar genetic constitution, were completely sterile (Wu, Johnson, and Palopoli, 1998).

Nevertheless, the validity of Haldane's Rule has not been entirely disproved by these observations and experiments. It is important to consider that the three phenotypic consequences of postzygotic hybrid incompatibility are controlled by different genes. These genes are not the same genes that regulate vitality, the function of the male sexual organs and the function of the female sexual organs. All three differentiation processes are regulated by different gene clusters and, therefore, do not necessarily have to be equally affected by X-autosomal imbalance. Indeed, it was substantiated that, especially the genes, which control male spermatogenesis, are subject to a specific rapid evolutionary speed. The genes for male fertility are said to be subject to a ten times faster evolutionary speed than the genes for female fertility (Wu, Johnson, and Palopoli, 1998). Moreover, the genes of male fertility change more rapidly than the genes for which both sexes depend on their vitality.

Nevertheless, biologists know little about the genes that cause postzygotic allelic incompatibilities, which are, in many cases, the first cause of speciation (Orr, 2009). Do few or many genes mutate, when an organism adapts to a new environment? Can the respective genes be identified? Are the same genes involved in adaptations to a new environment, if such adaptations occur several times independently from each other in different populations?

In recent years, approximately half a dozen genes causing sterility or a decrease in vitality in hybrids have been identified by evolution geneticists. These genes usually have basic and entirely different tasks, which at first appear to have nothing to do with regulating vitality or fertility. Some of these genes code for enzymes, others for structural proteins, and some even produce proteins that bind to the DNA and perform functions there. All of these genes, however, are distinguished by the fact that they have diverged extremely quickly in evolution (Orr, 2009). In *Drosophila*, a gene of the nuclear pore complex has been identified as the cause of hybrid sterility. The nuclear pore complex is a structure in the membrane of the cell nucleus and serves as a checkpoint in the process of the channeling in and out of macromolecules. The nuclear pore complex is composed of many proteins that coevolve at high speed. Thus, there is fast incompatibility if the parents stem from different gene pools (Tautz, 2009).

6.17 Sympatric and Allopatric Speciation

There are two different basic possibilities for why two organisms are not able to successfully crossbreed with each other. Either they live together and there exists a prezygotic barrier between them (see above) that prevents a successful zygote formation, or they live at different locations and, thus, cannot meet at all. The first scenario is called a sympatric distribution and the second scenario is called an allopatric distribution of the populations.

Differences between the two forms of separation are enormous and cannot be overvalued. The origin of two new species under sympatric conditions is different from the origin of two new species under allopatric conditions, although both processes are designated with the same term: speciation.

If two organisms live sympatrically and, therefore, meet each other regularly, they must evolve intrinsic properties that prevent mating. Completely different is the situation with two organisms that live allopatrically. Those organisms cannot meet each other for external reasons. Therefore, any prezygotic crossing barriers do not, in fact, have to exist. Allopatrically distributed organisms do not need traits that prevent zygote formation, which would not make sense biologically. Their mutual mating is prevented only by external limits, and these limits are not properties that the organisms possess themselves. Such external barriers are often, but not always, of a geographical nature. They can be oceans, mountain ranges, rivers or only highways that prevent an encounter of the separately living organisms.

In the case of certain Weevil beetles or nematode worms, however, allopatric conditions can also be produced by a lifelong confinement in the interior of a host plant that prevents a mutual encounter with each other, even in the same geographic position (McCoy, 2003). These animals spend their entire life cycles on or even in the interior of a single food plant without ever leaving the host plant. These examples have an external cause for separation that has nothing to do with a geographic separation. This form of separation is also an allopatry, although not through a geographical barrier. In every case of allopatry, however, the absence of crossbreeding is not based on the properties of the organisms themselves but, instead, is based on external barriers. Thus, the organisms themselves then carry no properties of separation at all within them. Their separation is not based on speciation genes. Those genes can, at best, evolve by pure chance.

Sympatric and allopatric separations are two entirely different evolutionary ways for speciation, and the question arises as to whether both types of group separation should receive the same name and whether both should be called species:

1) sympatric speciation is the origin of separated groups that are in competition and are separated by a control of selection at the same location, while
2) allopatric speciation is the formation of species by chance due to genetic drift at locations that are isolated from each other.

Sympatric speciation can occur if the sexual partners evolve new preferences for partner recognition. Females with newly evolved demands for specific traits in their male partners and the respective males, which conform to these demands, can split off as a distinct species and can separate from the rest of the members of the original species. This scenario is sympatric speciation resulting from sexual preference.

An additional possibility for sympatric speciation is the segregation of a few organisms into a new ecological niche, which opens up new resources and, thus, signifies an advantageous adaptation, for example, if a group of a monophagous beetle species switches to a new food plant and becomes adapted to this new food plant. For this separation to become stable, a corresponding selective partner choice must coevolve at the same time. Only under this condition, the conquest of the new ecological niche is not reversed again by backcrossing with the organisms that still have a preference for the former food plant. If, therefore, a few individuals within species *A* conquer a new food plant and rely on this new plant in the future, then they can only turn into the new species *B* if, simultaneously, an assortative mating is guaranteed that makes certain that *B*-organisms from now on mate only with *B*-organisms and not with *A*-organisms any more (Dres and Mallet, 2002).

Sympatric speciation means that the force of gene flow among the organisms of an established gene-flow community to again and again homogenize the genomes among the different individuals is overcome by other forces, such as (1) ecological adaptation and (2) sexual selection.

1) New niches offer a selective advantage to a new founder population, if the old habitats are completely occupied.
2) However, if a new resource is populated, only then is there a selective advantage, whereby the organisms find and select sexual partners with the same type of adaptation. Because the partner choice depends on specific signals, the conquest of new food resources together with newly developed signals for mutual partner recognition can immediately split a gene pool.

Sympatric speciation is exactly the opposite of the allopatric paradigm because selection is the driving force and not the contingency of an allopatric separation. In the case of sympatric speciation, adaptation to specific habitats, together with the erection of crossing barriers are subject to a high selective pressure, to prevent the remixing of the separating populations. Because of the importance of the role that selection plays, sympatric speciation is true Darwinism (Tautz, 2009).

Allopatric speciation, in contrast, is pure coincidence; it is simply the result of genetic drift. The theory of allopatric speciation is based on a twofold coincidence.

First, it is based on the assumption of a coincidental separation of two populations caused by external forces, such as, for example, geological forces or climatic events, and second, it is based on the assumption that, in the course of the allopatric lifetimes, mutations that lead to genetic incompatibility accumulate coincidentally. Neither of these two coincidences is founded on a quantifiable scientific theory. The allopatric paradigm is nothing but the putting-into-words of an *ad hoc* concept (Tautz, 2009).

If allopatrically separated populations become different, then selection has no "interest" in erecting prezygotic crossing barriers because the populations, which have become different, do not encounter each other anyway. For this reason, allopatric speciation usually needs a long time (Orr, 2009), while sympatric speciation can be a matter of only a few generations because it is stimulated by positive selection. If the allopatrically separated populations were to meet again later, then the differences that have originated by chance in allopatry can be an obstacle to remating. However, these differences do not necessarily have to be an obstacle.

Mayr has defended allopatric speciation as the only way in which species can originate (Mayr, 1963). The theory that speciation must almost always be allopatric has consequences that cannot readily be brought into agreement with our current knowledge of speciation. For example, according to the allopatric paradigm of speciation, those populations that live sympatrically without an external separation cannot split and, hence, would have to remain in existence "eternally." As long as speciation must almost always be allopatric, a population can never undergo speciation under sympatric conditions because, without geographical separation, there can be no split. (In this consideration, anagenetic alterations of traits are, of course, not counted as speciation; see Chapter 7).

Furthermore, the theory of allopatric speciation does not explain why there is rapid speciation in some groups of animals but not in other groups of animals, even though the groups should, in principle, have the same chance of being torn apart into isolated groups by external factors. For example, over approximately ten to a hundred thousand years, the Cichlides of the African lakes have produced, in every lake in which they occur, ten to several hundred species, while the members of other families of fish in the same lakes did not (Verheyen *et al.*, 2003) (see below). From this scenario, it clearly follows that the ability of a population to speciate also depends on the internal genetic factors of the organism itself, and these factors are specific to the respective groups of animals. In these cases, speciation is definitively not based on a separation by external factors but instead is based on the genetic constitution that is present, to a varying extent, in the different groups of animals.

Additionally, the paradigm of allopatric speciation does not explain why the number of species is so large in beetles even though most beetles are able to fly and, therefore, could easily overcome many external barriers. The (apparently) very large number of beetle species, however, must be acknowledged with some care. Currently, the order of Coleoptera (beetles) is divided into approximately 400 000 species. This order would be the order with the highest number of species in the animal kingdom except that the majority of these species are delimited by typological trait differences, which have often been determined in only a small number of specimens. Thus, it cannot be ruled out that, in the case of several so-called species, the trait differences found between those few specimens are only differences in the frequency distribution of specific alleles, which are the same in supposedly different species but differ only in their relative quantities. Thus, in some cases, geographically distant groups that are diagnosed by different traits might not be species; they could be only local populations or races, such as is the case with different human populations that are categorized by different allelic frequency distributions in their blood groups. Alternatively, they could be morphs, for example in the case of varying phenotypes of the Lepidoptera *Papilio dardanus* or *Zygaena ephialtes* (Color Plate 4) (Chapter 5). Delimitations with respect to gene flow barriers have not been investigated between most beetle species. Thus, it cannot be ruled out that the current number of 400 000 beetle species is a wrong number.

In retrospect, it is difficult to understand why sympatric speciation is a relatively recent discovery and was not recognized several decades ago. The biological processes that make sympatric speciation possible were sufficiently known in the former century. Sexual selection of partners (female choice) and monophagy, the specialization of the organisms of a species to eat a single food plant, have been known for more than a hundred years, and from these two processes, sympatric speciation can almost inevitably be postulated. Both phenomena make it easily to imagine that a slight alteration of traits within a population could lead to the segregation of a new population and, through this process, to speciation. A high percentage of insect species is monophagous. In tropical beetles and butterflies, the estimations come to 20–50% of the species (Schilthuizen, 2001). In temperate zones, many beetles (particularly weevils *Curculionidae*) are monophagous (Dres and Mallet, 2002).

Ernst Mayr did not give this concept any special importance because he was not an entomologist; he was an ornithologist. Because Ernst Mayr dominated the scientific position on this question for decades in the second half of the twentieth century, the difference between sympatric and allopatric speciation was given only a very subordinate importance (Mayr, 2000). Mayr defended the view that allopatric speciation would be almost the only process of speciation. He insisted in this belief because he was convinced that the species is a community of organisms that all are able to reproduce with each other, and that the species, therefore, is a homogeneously connected gene pool. This scenario meant that only very strong forces would be able to tear apart a gene pool. From this concept, the allopatric paradigm of speciation resulted.

The uncompromising dogma of allopatric speciation denotes a "Dark Age" of speciation research. It is based on few facts and mainly on authorities who dominated the field (Tautz, 2009). Modern evolutionary biology, however, demands quantifiable, testable models, in which the parameters, such as mutation rates, selection coefficients and migration rates, are measurable. The modern view has given priority to sympatric speciation because of empirical data. In doing so, modern evolutionary research approaches Darwin's idea again, after an age of "apostasy." Speciation is no longer first and foremost seen as drift, meaning coincidence without positive selection, but instead is seen as an "intended" process of evolution.

6.18
Sympatric Speciation in the Fruit fly *Rhagoletis*, in Cichlids and in the Fire Salamander

Mayr was confronted with the first experimental results on sympatric speciation by his own doctoral student. The student had been given the task of disproving sympatric speciation; however, to Mayr's discontent, he proved exactly the opposite (Schilthuizen, 2001).

The larva of the fruit fly *Rhagoletis pomonella* lives in North America on the berries of the Hawthorn (*Crataegus* spec.). In this region, there exist, however, also a few populations living on the Apple (*Malus* spec.). Because Apples do not occur natively in North America, but instead were introduced from Europe, the suspicion arose that the *Rhagoletis* larvae could have secondarily converted from the Hawthorn to the Apple and have formed a new distinct species on the new food plant.

Mayr put his doctoral student Bush on this topic, to prove that the differentiation between "Hawthorn flies" and "Apple flies" evolved through a geographical separation, in other words, under allopatric conditions, and that the two populations then only secondarily populated overlapping geographic areas. The resistance against the assumption that the change of food plant might have been speciation under sympatric conditions was mostly founded on the idea that the preference for a new food plant certainly was not capable of splitting up an existing population of insects because the flies with the new food preference would still merge with the flies of the old population.

Bush, however, found evidence that the flies not only had chosen a new food plant but also had used this new plant as the only place of courtship display and egg deposition. He noticed that the male's courtship display, which in the end leads to the females agreeing to mate, happened exclusively on those fruits, which then provided food for the larvae. The "Hawthorn flies" were courting on the Hawthorn, and the "Apple flies" were courting on the fruits of the apple tree. Bush subsequently had the idea that the fly's transition from the Hawthorn to the Apple was not solely a matter of larval food but was also a matter of secluded places for reproduction. Accordingly, an effective reproductive isolation would be based on only a few coevolutionary trait alterations. Thus, Bush made the exact construct probable that Mayr had expected and hoped he would disprove (Schilthuizen, 2001).

The final proof that sympatric speciation indeed occurs was still some time in coming, however. Nature provides only a few undoubted examples because it is almost always possible that alternative allopatric scenarios can still be revealed underneath. There are conditions that must be substantiated to state that two populations that currently live in sympatry really have diverged under these sympatric conditions. This scenario would require showing that, in the course of divergence, an encounter of the members of the separating species was not prevented by external geographic or similar barriers.

This verification became possible in a small crater lake in West Cameroon that has a surface area of only half a square kilometer. According to geological calculations, this lake is determined to have originated only 5000 years ago. In this lake, there are five endemic Cichlid species of the genus *Tilapia*. They stem from a mother species, which lives in neighboring rivers. Two of these five species have been compared to each other in more detail (Schliewen *et al.*, 2001). One species preferably lives close to the shore and consumes slightly different food than the second species, which lives more in the open water. However, both occurrences overlap. It is concluded that this situation is an example of sympatric speciation, which has happened endemically in this lake within at most a few hundred generations. Different ecological adaptations of the fishes together with assortative mating have split the gene flow cohesion of the original combined species.

Another example for sympatric speciation through local adaptation is presented by the Fire Salamander (*Salamandra salamandra*) in Germany. Fire Salamanders constitute a special case among the amphibians that are native in Central Europe. While most of the amphibians frequent ponds and pools for a specific time period in the spring, to mate and deposit eggs, Fire Salamanders exclusively mate on land. The Fire Salamander is viviparous. The females migrate alone to the bodies of water in the spring, to deposit their larvae there. Usually, the larvae are deposited in small flowing waters, where they grow up until the completion of metamorphosis. In the forest regions of Western Germany, however, there are a few populations in which the females do not deposit the larvae in flowing waters, but instead deposit them in small ponds with standing water. The larvae, which grow up here, must, in several respects, be differently adapted to their habitat than the larvae of the flowing waters. Their food supply is different, the oxygen content of the water is different, and because of the danger of dehydration, their development is more rapid than that of the larvae of the flowing water population.

This scenario either indicates a large plasticity for the larvae, or the salamanders of the flowing and standing waters are examples of two different genetic adaptations. Experiments with mutually translocated larvae show that the translocated animals retain the behavioral patterns of their original populations, indicating a genetic basis (Weitere *et al.*, 2004). Thus, the existence of two different populations is not a simple, vivid answer to different environmental conditions. Because the mating of the adult animals occurs in the common habitat outside of the bodies of water on land and without proximity to the separated larval habitats, there must be reproductive barriers that prevent the mating among the individuals of the different genotypes.

"Flowing water ecotypes" and "pond ecotypes" trace back to a common founder population, which recolonized Central Europe after the last ice age. Afterward, both populations have separated and today coexist in forest areas of various regions in Germany. Flowing water salamanders are all genetically similar to each other at distant locations in Western Germany. The pond salamanders, on the other hand, are also all genetically similar to each other, independent of the geographic location. However, the flowing water and pond salamanders differ significantly from each other, even at the same location. Accordingly, there is more gene flow per unit distance among the flowing water salamanders at different locations as well as among the pond salamanders at different locations compared to different "ecotypes" at the same location (Weitere *et al.*, 2004).

6.19
Reproductive Incompatibility is Different than Phylogenetic Distance

Reproductive incompatibility is not the same as phylogenetic distance. Therefore, both criteria cannot be used for a common species concept. A species that, as a group, has an evolutionary distance to another species is not the same as a species that, as a group, is reproductively isolated from another species. In other words, the phylogenetic species is not the gene-flow community, at least not as long as the terms are applied to evolutionarily young species.

The phylogenetic distance of two organisms and their mutual genetic compatibility must not be lumped together, although a long-lasting phylogenetic separation in several cases has resulted in genetic incompatibility. However, there are evolutionarily and genetically far distant organisms that are still reproductively compatible with each other (Lande, 1980; Coyne and Orr, 1997). Vice versa, there are organisms that separated from each other only a short time ago, which could differ less genetically than humans of different populations but which are reproductively well-separated from each other. Among these organisms are the Cichlids in many African lakes (Schliewen *et al.*, 2001; Meyer, 1993). Kinship and sexual compatibility are not the same thing. The action of defining a species to be a group that has exceeded a minimum of phylogenetic distance does not represent the species concept of the gene-flow community (see the criticism on barcoding in Chapter 4).

In contrast to expectation, the occurrence of hybrids between different species is not always a consequence of a close kinship of the parental species. The concept that two species would not be able to combine with each other, and if they were able, then they simply would not be "real" species yet because they were still too closely related to each other, is not sustainable in this simple form. There are evolutionarily young species (meaning closely related species) that are cleanly separated and that hardly hybridize with each other. Indeed, only a single or a few gene mutations are sufficient to raise reproductive species barriers (see below) (Phadnis and Orr, 2008; Prud'homme *et al.*, 2006).

That the equation of phylogenetic distance and reproductive isolation is not justified follows already from the difference between plants and animals. Among other reasons, plants differ from animals in that they break species barriers much more often than animals, and they hybridize with each other. However, the ability of plant species to form species hybrids much more often than animals is not related to the fact that plant species are more closely related to each other than animal species. The same consideration applies to bacteria (see above). It is not the phylogenetic proximity of plants to each other that causes them to be able to hybridize; instead, it is the plants' different biology that makes them have the tendency to have more species hybridization than animals.

The commonly found equation of phylogenetic distance and reproductive isolation results from the outdated assumption that the gene flow within a species is always strong and extensive, so that de novo evolution of allelic mutants either disappears rapidly or quickly spreads across all organisms of the gene-flow community (Dobzhansky, 1937; Mayr, 1942). This phenomenon is called "fixation" of the alleles.

However, in a geographically widely spread species, gene flow can strongly decrease with distance. It may be possible that many de novo mutated alleles never bridge the distance between populations of a species that are geographically far apart (isolation by distance, see above).

6.20 Phylogenetic Distance and Reproductive Incompatibility in Two Species Pairs, Polar Bear (*Ursus maritimus*) and Brown Bear (*U. arctos*), in Comparison to Grey Wolf (*Canis lupus*) and Coyote (*C. latrans*)

The Polar Bear (*Ursus maritimus*) is a relatively young species, which split off of a population of Brown Bears (*Ursus arctos*) living in north-east Siberia approximately 200 000 years ago (Breiter, 2008). The usual classification into two species, Polar Bear and Brown Bear, has been conducted according to phenotypic trait differences and does not correspond to descent relations. Brown Bears are widely spread across the northern Holarctic and occur southwards as far as Spain, Italy, Persia and India. They are distributed across several separate territories.

Some populations of Brown Bears have long since been isolated and are phylogenetically further apart from each other than the Polar Bear is from the northeastern Siberian Brown Bear (Talbot and Shields, 1996). The term "Brown Bear" does not correspond to a species because it is neither a natural gene-flow community nor is it a monophylum. Instead, Brown Bears is a collective term for several separate gene-flow communities that have a different evolutionary relation towards each other. Retaining the traditional term "Brown Bear" as an artificial species for pragmatic reasons would mean that this group would be a paraphylum (Chapter 7) because the Polar Bear would have to belong to it also.

Brown Bears are only classified into a common species for subjective reasons because of their brown fur, their relatively long ears and their terrestrial way of living. As is easily apparent, the Polar Bear looks different, which is only an evolutionarily young adaptation to its life in ice and water; however, these are a small number of traits that are outwardly visible to humans and do not determine any information about kinship.

Despite the close kinship of the Polar Bear and the northeastern Siberian Brown Bear, hybrids are very rare, although they are fertile if produced in captivity (Breiter, 2008). At least in the case of the female partner being the Polar Bear, it is evident why hybrids are so rare. The significantly smaller Brown Bears are apparently unable to compete with the male Polar Bears for mating rights because the female Polar Bears exhibit a distinctive preference for large partners, and the larger partners are solely the Polar Bears. Thus, it is not the phylogenetic distance between the Brown Bear and the Polar Bear that makes hybrids among them so rare; instead, it is an ethological barrier that prevents mating.

In contrast to the Bears is the relation between the Grey Wolf (*Canis lupus*) and the Coyote (*C. latrans*) in North America. The Grey Wolf and Coyote separated roughly two million years ago, which is a ten times greater phylogenetic distance than between the Polar Bear and the Brown Bear (Wayne, 1993). Nevertheless, hybridizations frequently occur between the Grey Wolf and Coyote. While the Coyote was originally native in the south of the United States, it successfully spread to the north in the last 100 years and has penetrated into most of the Grey Wolf's living habitats. Here, successful hybridizations frequently occur; as a result, most of the North American Grey Wolves already contain Coyote genes. In the province of Quebec in Canada, almost 100% of the Grey Wolves contain portions of Coyote DNA in their genome (Wayne, 1993). Apparently, strict speciation genes are missing that could prevent a hybridization between the Grey Wolf and the Coyote. It would be faulty to infer a close kinship from the frequency of hybridizations between the Grey Wolf and the Coyote. Furthermore, disregarding one exception (the so-called Red Wolf; see Chapter 2), there are no pronounced hybrid populations, but the animals are still distinctly recognizable as Grey Wolves or as Coyotes in spite of all of the blending.

From these and other examples, it follows that the breaching of species boundaries is not directly correlated with the degree of phylogenetic distance. Furthermore, there is no universally valid biological law that would prove the idea that a considerable percentage of mixed matings between two species must necessarily lead to the blending of these species (and thus to the disappearance of one of the two species), even though it is often assumed to be the case. As long as the percentage of hybrids remains constant (and does not increase in an ongoing fashion) the continuance of the species is not endangered, and species hybridizations can be tolerated over long evolutionary time frames (see the example of the black European Flycatchers in Chapter 4). Apparently, some species can "tolerate" well that a significant percentage of their conspecifics enter mixed matings and produce hybrids, while other species do not "tolerate" this occurrence, which, in those cases, has the consequence that one of the two species can be exterminated by hybridization (see the example of the Ruddy Ducks further below).

6.21
The Herring Gull (*Larus argentatus*) and the Greenish Warbler (*Phylloscopus trochiloides*), a False and a True Model for the Ring Species

As mentioned above, the individuals from geographically distant populations of a species in many cases cannot be successfully crossed with each other (Figure 6.2a). This scenario is termed isolation by distance. A special case of this phenomenon is the ring species. A ring species is a group of populations that show the normal phenomenon of isolation by distance between their very distant individuals; however, they exhibit the exceptional phenomenon that those distant individuals encounter each other under natural conditions (Figure 6.2b).

This scenario can happen in those cases in which the different populations of a species surround an inhospitable geographic region, in which they cannot live. This region could be a mountain massif, an ocean or the polar region. The different populations of the species surround this inhospitable region until the expansion is closed at a geographic area where the populations encounter each other at both ends of the expansion. In the end, they form a ring-shaped distribution zone. Whereas all of the adjacent populations are reproductively connected to each other, the most distant populations lose their reproductive compatibility when they encounter each other at both ends of the circle (Figure 6.2b). They are no longer able to successfully crossbreed under the natural conditions present there. Isolation by distance has more and more estranged the populations over time, during the expansion. With the example of such a geographic ring-shaped chain of populations, surrounding the polar region or other inhospitable regions such as mountains or deserts, the term "Rassenkreis" (racial circle) was originally introduced (Rensch, 1947). This name is a correct name; however, it has not become established terminology. It has been replaced by the less appropriate term "ring species."

For a long time, the Herring Gull (*Larus argentatus*) was considered to be the classic example of a ring species. The following model was stated: The Herring Gull originated in southwestern Asia, and from there, it spread through northern Asia and continued through North America, until it finally migrated across the North Atlantic Ocean to penetrate Europe. Because of this scenario, the Herring Gull was said to be the prime example for a ring species that has spread around the North Pole.

According to Mayr (1942), the Herring Gull originated in the Aral-Caspian region and spread through Mongolia to northeastern Siberia. The local race of northeastern Siberia received the distinct species name *Larus vegae*. The expansion continued to North America, where the Herring Gull was designated as *L. smithsonianus*. In North America, probably caused by the ice age, a new "race" *L. argentatus* was separated from *L. smithsonianus*, which then immigrated into Europe after the ice age. In Europe, *L. argentatus* encountered the Lesser Black-backed Gull (*L. fuscus*), which was originally native in the Baltic region. Here, at the hypothetical end point of the circumpolar expansion ring, the two gull species do not mate with each other. From west to east in Eurasia and North America, all of the adjacent races are reproductively compatible with each other, while the two end populations of the expansion ring (the Herring Gull and the Lesser Black-backed Gull) have reached full reproductive isolation and coexist in Europe similar to distinct species alongside each other.

The Herring-Gull example has been the standard model for a ring species for more than half a century and, as such, has entered every text book. However, recent investigations have not been able to confirm this orderly model (Liebers, de Knijff, and Helbig, 2004). A comparison of mitochondrial DNA sequences could not support a key element of the ring species hypothesis, namely the postulated close relation of the North-American *L. smithsonianus* to the European *L. argentatus*. The haplotype sequences that are distinctive for the Nearctic *L. smithsonianus* were not rediscovered in European *L. argentatus* gulls, not even in Iceland. There are no indications for an evolutionarily young bifurcation between *L. smithsonianus* and *L. argentatus*.

Thus, the Herring-Gull complex is not a ring species. The available data do not support the conception of a ring-shaped expansion. Instead, this case is a case of diverging expansion waves that emerged from two separate glacial refuges. The mitochondrial data speak for an expansion from a continental Eurasian retreat area, on the one hand, and a second retreat from a North Atlantic refuge.

Unlike the Herring Gull, a valid example for a ring species is the Greenish Warbler (*Phylloscopus trochiloides*). This species forms a racial circle in Central Asia (Irwin, Bensch, and Price, 2001; Irwin *et al.*, 2005). Climate data and the comparison of mitochondrial microsatellite sequences permit us to reconstruct that the animals spread from the south of the Tibetan Plateau to both the western side of the plateau and the eastern side around the entire Tibetan Plateau. The Tibetan Plateau is, today, surrounded by several races of Greenish Warblers, which are all reproductively connected to each other. The molecular data show that gene flow occurs between the neighboring races. On the northern side of the Tibetan Plateau, however, the races from the west and the east encounter each other; here, they are mutually genetically incompatible. Between the two races in the contact zone, gene exchanges no longer occur. Thus, the Greenish Warbler, as established, forms a real ring species.

A ring species is often considered to be an exceptional peculiarity. However, this idea is not correct. A ring species is exceptional only with respect to the fact that distant individuals of a species encounter each other under natural conditions. Otherwise, a ring species is nothing other than the normal phenomenon of isolation by distance, the phenomenon that the individuals from geographically distant populations of a species in many cases cannot be successfully crossed with each other.

A ring species is often considered to be the start of a speciation, as though a species would be on the brink of bifurcating into two separate species. However, this conclusion is not mandatory because there is no necessity of assuming that the ring species is not stable as a species and would shortly have completely fallen apart into two or more species (Irwin, Bensch, and Price, 2001; Irwin *et al.*, 2005). No natural law compels us to consider the coexistence of reproductively compatible and incompatible populations within the same species as an evolutionarily short-lived transitional process. This state can very well be stable. Only the purely human need to set rigid class boundaries converts a ring species into a transitional phenomenon.

6.22 Allopatrically Separated Populations are Always Different Species

The examples of isolation by distance and the example of the ring species clearly show that it is not possible to define a species as a community of organisms that are all cross-fertile. Instead, the mutual cross-fertility of the organisms of a species in some cases is restricted to adjacent populations only. The cohesion of the individuals within a species must be understood by stepwise allele exchange through intermediate populations. With increasing distance, the potency of the organisms of a species for successfully interbreeding often decreases gradually (Ford, 1954; Baldwin *et al.*, 2010).

However, although the distant individuals of a species could lose their mutual ability to be crossed over large geographic distances, each population is connected with at least an adjacent population. This scenario distinguishes isolation by distance and distinguishes the ring species from allopatry. In the case of isolation by distance, no population is completely isolated. In contrast, allopatry means that the connection is lost. Allopatry is the split of one gene-flow community into two separate gene-flow communities. Understanding the species as a gene-flow community instead of as a reproductive community has extensive consequences for the species status of allopatrically separated populations. Allopatrically separated organisms cannot encounter each other. No alleles can migrate from one of the communities into the other. The cohesion of the two groups no longer exists, and therefore there is no logic justification to join allopatrically separated populations into a common species.

Understanding the species as a gene-flow community means that allopatry always is speciation, independent of whether the allopatrically separated populations have maintained their putative interbreeding potency. Because the phenomenon of isolation by distance documents that distant organisms of a species often lose their mutual cross-fertility, interbreeding potency cannot be a species criterion. If two allopatrically separated gene-flow communities later come together again, encountering each other, they could revert into a single species, as before. However, as long as they are separated, they are different species.

Opponents of the view that allopatry means speciation must consider that all of the other ways to solve the problem of allopatry are not free of contradictions. Mayr attempted to solve the problem of allopatry by introducing the concept of the *potential* ability to crossbreed. Hidden behind this concept, however, are several inconsistencies. Mayr states that the organisms of allopatrically separated populations would still belong to one and the same species in every instance in which they would successfully interbreed, given the chance. "Species are groups of actually or *potentially* interbreeding natural populations that are reproductively isolated from other such groups" (Mayr, 1942).

However, the problem of allopatry is not solved by this consideration. There are practical as well as theoretical objections against the concept of a "potential" ability to crossbreed.

1) First, the practical objection:

   The experimental proof of a potentially successful crossing is very difficult and cannot be performed at all in most cases because the behavioral biology of partner identification and courtship is very complicated in nature. Under experimental conditions, the behavior of most animals is significantly disturbed, so that it is difficult to decide whether there are prezygotic barriers. In many animals, the final successful mating requires a delicate and sensible interaction of mutually attuned behavioral patterns that are susceptible to failure. Males and females react specifically to each other with optical, acoustical or olfactory signals and, thus, distinguish themselves from alien partners and execute a mating ritual, which often leads to a successful copulation only after several subsequent rites. Only the strict obedience of these prezygotic mating rites warrants a species-specific pairing.

   These complex mating rituals are, in most cases, very difficult to reconstruct under the artificial conditions of captivity. Behavior disorders and sexual starvation in cages or zoos often result in matings that occur fairly arbitrarily. It is tremendously difficult and, in many cases, hardly ever feasible to reconstruct the complicated mating behavior that leads to prezygotic compatibility or incompatibility. The concept of the "potential" ability to crossbreed has a completely different meaning in a laboratory or in a zoo than in the wild. Under artificial conditions, many species interbreed that would not do so in the wild. In some cases, one needs only to lock two alien species into a common cage to obtain interspecific hybrids (Wirtz, 2000), for example, crossings between a lion and tiger are not difficult to obtain in a zoo. Surprisingly, such "slips" are often successful and lead to vital and absolutely fertile offspring. Selection has not stopped the begetting of hybrids such as these because they would not occur in the wild. As a consequence, the phenomenon of the "potential" ability to crossbreed, while it is nicely expressed in words, cannot be verified experimentally in most cases.

2) The theoretical objection against a species concept that includes the potential ability to crossbreed is the following:

   If the species would be a truly existing object and not an artificially constructed class of objects that share some trait similarities, then a species is an individual (Chapter 3). Species as individuals, as philosophically analyzed in detail by Ghiselin (1997), are units of relationally connected organisms. They are historically singular occurrences that have, at one point in time, originated, and will become extinct later, at another point in time.

   Nevertheless, species as individuals cannot be allopatrically split. The connection of cohesively linked organisms of a species requires an actual gene flow, not a "potential" ability to crossbreed. An individual cannot consist of members that physically separated and are only potentially connected. Potentially cross-fertile organisms are not parts of an entity that exits in reality.

3) However, even if this argumentation is to be ignored, the phenomenon of isolation by distance already makes the concept of the conspecificity of potentially cross-fertile organisms useless.

   Because, in several cases, as a consequence of isolation by distance, far distant organisms could lose their potency to interbreed anyway, in spite of belonging to the same gene-flow community, it appears to be inconsequential to determine the species status of an individual organism of a species by its potency to interbreed with another arbitrarily chosen individual of this species. The term "potential cross-fertility" has become rather meaningless in the light of the phenomenon of isolation by distance.

   Therefore, the potency of successful interbreeding has also lost its viability as a test for the species status of allopatrically distributed organisms. Allopatrically separated organisms cannot be considered to be conspecific, even if they can be presumed to have the potency to interbreed.

6.23
Species Hybrids as Exceptions without Evolutionary Consequences

Hybrids between different species probably occur among many species, as long as the species are phylogenetically not too far apart from each other. There are species hybrids and genus hybrids. However, family hybrids are probably impossible in higher animals. Not impossible are the hybrids between different species within families, for example, within Canines (*Canidae*) or within Felines (*Felidae*); however, hybrids between dogs and cats are probably impossible. The occurrence of hybrids is, therefore, without importance for the taxonomical classification of species. The position that the occurrence of hybrids determines that these are not "real" species elevates the category of species to the hierarchical level of family. The occurrence of hybrids is at best a criterion for family status.

Thus, the occasional occurrence of hybrids is unimportant taxonomically. What is crucial is how often those hybrids occur. The taxonomist should not ask the question of whether hybridizations occur and whether those hybrids are then fertile or sterile. It does not greatly matter that hybridizations occur. Only if a considerable percentage of species hybridizations occur is this concept of importance for the population-genetic structure and for taxonomical classification. In most cases, the (prezygotic) assortative mating predominates, which keeps the number of hybrids within a limit. If the sexual partner of one's own species is preferred to one of an alien species, the system could be stable, despite the occurrence of species hybrids. In the extreme case, however, hybridization can lead to the extinction of a species (see below). Furthermore, it matters whether the hybrids are (postzygotically) disadvantaged with regard to their vitality and fertility in comparison to the purebred offspring. This disadvantage does not always occur. The textbook example of the Horse and Donkey, for which the hybrids are almost always sterile, is not a very representative example.

Hybrids between different species of plants have always been a familiar phenomenon to botanists (Ehrendorfer, 1984; Gornall, 2009). Many plant species, especially perennials and wooden plants, form species hybrids, which are also often 100% fertile. In the genus *Salix* (willow), a hybrid was artificially produced through the sexual crossings of 13 different species (Gornall, 2009). In *Rosaceae* and *Poaceae*, genus hybrids are common, too, although in the case of genus hybrids, the question arises as to whether the bestowal of the genus rank was truly justified for those species. In the flora of Great Britain, 780 species hybrids are listed along with 2500 plant species (Gornall, 2009). The top rank among the plant hybrids is occupied by the orchids, for which, at least within the genera, close to all species hybridize with each other. The paternity analysis of a cultured form of orchids exposed this case as the product of eight different genera. Only in most species of the orchid genus *Ophrys* hybridization hardly ever occurs because the species of this genus are pollinated by very specific insect species. These insects are the Scoliid Wasps, Sphecoid Wasps and representatives of many other families of *Hymenoptera*. Each pollinator species recognizes only a single *Ophrys* species and is, thus, responsible for the fact that the pollen cannot be transferred between different *Ophrys* species (Paulus and Gack, 1983).

In animals, species hybrids are rarer than in plants but also occur among most related species. Hybrids are more common than was assumed in the past. Recently, the "Handbook of avian hybrids of the world" has been published; on 608 pages, it describes the worldwide occurrence of thousands of hybrids in birds (McCarthy,

2006). In contrast to expectations, these hybrids often do not have an intermediary phenotype. Because of the dominance of many genes, the phenotype of only one parent can predominate, with the result that the hybrid is not readily recognized because it resembles only one of the two parental species. This scenario is often the case in ducks (Randler, 2000).

An increased chance for species hybridization occurs when one of the sexual partners suffers sexual starvation. For example, this case occurs if one species faces extinction. In this case, members of the rare species are only rarely found by an individual looking for a sexual partner, while members of the other species exist in abundance and are, therefore, "in a pinch" accepted as sexual partners. The hybridization with the still frequent species is an additional threat for the rare species, which can lead to the extinction of the rare species (Frankham *et al.*, 2001). This scenario was, for example, observed in the case of a rare species of Fur Seals on Marion Island in the southern Indian Ocean (Wirtz, 2000). Species hybridizations also occur more often towards the end of the mating season, when the members of the same species are almost all paired up, while those of the other species are still available.

Especially impressive are the many hybridizations among almost all species of *Anatidae* (ducks, geese, swans). The *Anatidae* are so-to-speak the orchids among the animals (Randler, 2000). In an investigation by Scherer and Hilsberg (1982) on the hybridization of the *Anatidae*, 418 different hybrids originate from a total of 149 species; 52% of the hybrids even comprise genera:

Mallard (*Anas platyrhynchos*) with Egyptian Goose (*Alopochen aegyptiacus*).
Egyptian Goose (*Alopochen aegyptiacus*) with Greylag Goose (*Anser anser*).
Greylag Goose (*Anser anser*) with Mute Swan (*Cygnus olor*).
Swan Goose (*Anser cygnoides*) with Mute Swan (*Cygnus olor*).

Observations speak in favor of at least a part of the hybrids being fertile, so that the gene exchange between the species is further propagated. However, it is often not possible to pursue the question of whether the hybrids produced in nature are fertile. This difficulty occurs because the goslings that follow a couple of geese or the ducklings that follow a leading duck mother do not necessarily have to descend from those parents. In *Anatidae*, there are occurrences of exchanges, including nest robbery, unfaithfulness and adoptions.

6.24 The Example of Some Duck Species: Extinction through Hybridization

Nevertheless, the frequency of the occurrence of species hybrids should not contradict the fact that there are biological laws that prevent incidences of this type. The occurrence of reproductive isolation is still the central objective in the evolution of species because, if the organisms of one species would not be reproductively isolated from each other, then mixed types between the species would occur in an unlimited way. An infinite production of mixed types carries the danger of one of the parental species gradually becoming extinct (Korol *et al.*, 2000). This scenario would reduce biological diversity because precious adaptations to specific habitat conditions would become lost in this way. Adaptations to specific habitat conditions, however, are the crucial criterion of speciation, and the exact set of adaptations would become lost, again by hybridization.

As long as the species live allopatrically in separated regions, there is no danger of blending. As soon as one species enters the geographical range of the other species, however, this danger arises, if neither pre- nor postzygotic barriers are present. The blending between species can lead to the disappearance of one species, by the "defeated" species merging into the dominating species.

In this way, the Mallard (*Anas platyrhynchos*), which was introduced to Australia and New Zealand, has through hybridization in some areas completely displaced the originally native Pacific Black Duck (*Anas superciliosa*) (del Hoyo, Elliott, and Sargatal, 1992). Pacific Black Ducks occur only in Australia, New Zealand and New Guinea and on a few Indonesian islands. In New Zealand, the coexistence of the Mallard and the Pacific Black Duck has led to the fact that today only 15 to 20 percent of the total population is pure Pacific Black Ducks – compared to 95% in the year 1960 (Schäffer, 2004). If this hybridization continues further, the Pacific Black Duck will disappear.

In Southern Europe, the North-American Ruddy Duck (*Oxyura jamaicensis*) currently threatens the Mediterranean to Caspian White-headed Ruddy Duck (*Oxyura leucocephala*). These duck species are phenotypically fairly different. For a bird watcher, it is completely impossible to confuse them. In the 1940s, Sir Peter Scott, the founder of the Wildfowl and Wetlands Trust, imported only seven American Ruddy Ducks (four males and three females) from America to England, for the purpose of keeping the birds in his collection of waterfowl in Slimbridge, which is in south-western England. From there, between 1953 and 1973, approximately 90 descendants of those birds, in total, flew away and started breeding outdoors in England (del Hoyo, Elliott, and Sargatal, 1992; Schäffer, 2004). In the year 2000, already more than 5000 Ruddy Ducks were breeding in Great Britain. Today, the Ruddy Ducks have spread across many European countries as far as Spain, where they interbreed with the rare and highly endangered native White-headed Ducks. Apparently, no reproductive barriers exist, either postzygotic barriers or prezygotic barriers. Because, in the last half century, the White-headed Duck has severely decreased over the entire geographical range, this hybridization threatens the further existence of the White-headed Duck.

These examples also make clear that it is impossible to conclude from the phenotype of two species their species status. The Mallard and the Pacific Black Duck as well as the American Ruddy Duck and the Old World White-headed Ruddy Duck are diagnostically very different, so that they have been designated to be different species without any doubt. However, their intermixture after geographical immigration has clearly shown that there are (almost) no species barriers. A very different phenotype is not the same thing as the existence of species barriers (Chapter 4). Significantly, different looking allopatrically separated populations can still be mutually absolutely fertile, if they encounter each other.

These examples again make Mayr's concept of the species as a reproductive community questionable. If all of the individuals that are putatively cross-fertile with each other would belong to a common species, the Mallard and Pacific Black Duck as well as the American Ruddy Duck and Old World White-headed Ruddy Duck definitely would have to be considered as being conspecific. However, this action has not been taken; they can be found as different species in every bird identification book.

6.25 The Origin of Reproductive Isolation Through Reinforcement

Yet not in all cases does the encounter of allopatrically separated species carry the danger of the hybridogenic disappearance of a species. In an encounter of previously allopatrically separated species, there are two possibilities: Either the two species blend, which because of the dominance of one species almost always leads to the termination of the other species, or there are barriers, which keep the blending contained or entirely prevent it. If species are allopatrically separated, then it is not foreseeable which of the two possibilities will take place if the two species encounter each other again.

If two populations are allopatrically separated, then there is no selective pressure to evolve any type of species barrier. Prezygotic or postzygotic genetic incompatibilities only arise by chance. The emergence of both types of barriers is not promoted under allopatric conditions because there is no biological need for them; the individuals do not encounter each other anyway. Under allopatric separation postzygotic genetic incompatibilities usually have a higher chance to emerge than prezygotic incompatibilities, because they are not as specific as premating or fertilization barriers.

This line of reasoning has a remarkable consequence. If two allopatrically separated populations secondarily come together in the course of the immigration of one population into the domicile of the other population, then postzygotic mating barriers could exist, but no prezygotic mating barriers would exist. As a consequence, the two populations mate with each other. However, because of postzygotic genetic incompatibilities, the offspring of such mating is reduced in vitality and fertility. This effect is a significant disadvantage in the fitness of the parents that produce such hybrids because their reproduction will not be very successful. This outcome leads to a selective pressure that promotes the expression of genes for prezygotic incompatibility, for example aversion between the different sexual partners. This scenario then prevents mixed matings. Assortative mating has then evolved.

This phenomenon is called "reinforcement." Reinforcement is the protection from disadvantageous mating furthered by selection and the prelude to the development of prezygotic barriers in the encounter of populations that postzygotically already exhibit species barriers. Reinforcement serves to protect specific gene combinations from a collapse through genetic recombination (Korol *et al.*, 2000).

Of course, sexual partners cannot know beforehand how their produced offspring will look. However, the laws of evolution ensure that parental properties prevail that guarantee a correct partner choice. As a positive result, vital and fertile offspring are produced, which then also successfully inherit the prezygotic speciation genes of the parents. Then, if the offspring of a purebred mating have advantages, then selection also benefits those parental properties that guarantee a "correct" partner choice. The underlying selective pressure ensures that those traits, then, quickly prevail in most cases.

Reinforcement has been verified for many examples, as follows:

When, in the 1920s, the Lesser Black-backed Gull (*Larus fuscus*) penetrated from its native breeding area in the Baltic Sea into the geographical range of the Herring Gull (*Larus argentatus*) in the southern North-Sea region in Germany and Holland, both species at first mated relatively often. Hybrids of both sexes are fertile. Today, after only a few decades, the two species have almost completely stopped interbreeding with each other (Haffer, 1982). Stuffed hybrids can still be found in Dutch museums and are today considered a valuable rarity. Apparently, the hybrids had postzygotic disadvantages, although these are not known. Accordingly, because in the first contact a prezygotic isolation came into being (or strengthened), in the course of only a few generations and as a result of natural selection, there is a bias against hybrids.

A second known example is of *Drosophila*: At Mount Carmel in Israel, there is a gorge that ironically is called "Evolution Canyon" (Korol *et al.*, 2000). The northern and the southern slope of this canyon are only a few hundred meters apart from each other, but they differ drastically from each other microclimatically because of having a different solar radiation and humidity. On both canyon slopes, certain populations of *Drosophila melanogaster* have adapted to the strongly different microclimates, for example, with regard to the ground temperature that triggers egg deposition. Immediately after the evolution of these adaptations, a sexual blending of the populations of the opposing mountain slopes would be fatal for the preservation of the linkage of the newly evolved traits. The linkage of the new alleles would have been destroyed by genetic recombination. Thus, a strong selective pressure has ensured a quick coevolution of an assortative mating behavior. A strong assortative partner choice with a preference for only the flies of the same mountain slope evolved, while the organisms of the opposing slope were avoided. In doing so, the newly evolved gene pools were kept apart. This scenario implies the de novo origin of two species of *Drosophila* under sympatric conditions because the distance of only a few hundred meters between the two mountain slopes allowed for a daily encounter of different flies. During the process of species formation, there were no external, allopatric barriers.

Between the two extremes, there is no reinforcement, but there is apparently unrestricted blending (in the example of the Ruddy Ducks) and there is strong reinforcement with fast development of mating barriers in the example of the European North-Sea Gulls; there are stable intermediate solutions. There are species that hybridize frequently with each other. However, despite frequent hybridizations, the species do not lose their identity. This outcome results from the fact that the percentage of species hybrids does not increase in time in the long run. If this condition is satisfied, then the occurrence of frequent hybridization apparently does not constitute any danger for the continued existence of the parental species. In specific cases, a constant amount of blending appears to be bearable for a long time. In fact, blending can even be beneficial for one of the species (see the example of Darwin's Finches, further below).

One example for a stable species status despite frequent hybridization is the Eurasian species pair Carrion Crow (*Corvus corone*) and the Hooded Crow (*C. cornix*) (Chapter 5). The breeding areas of the two species overlap in a narrow belt with a width of less than 50 km, in which hybrid formation occurs frequently. This hybrid zone appears to be stable, with no indications that the occurrence of species hybrids would endanger the distinctness of the two well-separated phenotypes and genotypes (Haas and Brodin, 2005).

6.26 Hybridogenic Speciation

The extreme case among the examples, in which species barriers are, to a certain extent, open, which even has evolutionary importance, is hybridogenic speciation. This term is understood to be the origin of a new species through the hybridization of two separate parental species. At first glance, hybridogenic speciation appears to be a contradiction to the species concept of the gene-flow community because, according to this concept, the species are defined to be reproductively isolated from each other. Here, there is not an example of vague boundaries resulting from mutual gene introgression between two species, but it is an example of the origin of a new species because of the penetrability of the species boundaries of two parent species. Hybridogenic speciation is the origin of a new species, because species hybridization is furthered by positive selection.

Hybridogenic speciation should not be confused with the fusion of two formerly separated species into a new common species. This process leads to a decrease in species numbers, to species loss. Hybridogenic speciation, in contrast, does not lead to a decrease in species numbers but instead leads to the origin of a new species from two separate parental species, without these ending their existence. Hybridogenic speciation is the origin of three species from two species. In flowering plants, this type of scenario is, indeed, a common evolutionary process for the origin of new species. A total of 2–4% of all of the species of flowering plants are believed to have originated in this way (Turelli, Barton, and Coyne, 2001; Schluter, 2001). Extended to all plants, not only to flowering plants, even 11% of all of the species are said to have a hybridogenic origin (Barraclough and Nee, 2001). In animals, hybridogenic speciation is much rarer.

Why is hybridogenic speciation more common in plants than in animals? What is the reason for this difference between animals and plants? There are two different reasons. First, there are many more examples of self-fertilization in plants than in animals, and second, there are many more examples for tetraploidies in plants than in animals. Both differences are responsible for the relatively high frequency of hybridogenic speciation in plants compared to animals.

1) **Self-fertilization**: The first important difference between plants and animals is a plant's ability for self-fertilization, a property that occurs much more rarely in animals. Species hybrids would remain as evolutionarily insignificant and rare incidences if the hybrids on their own would not have the ability to rapidly erect a new population of numerous new individuals. Only a population that is rich in individuals would be able to compete with both parental species as a distinct new species in the struggle for life. Only this scenario gives the hybrids the chance to prevail and survive as a newly evolved species.

   Normally, species-hybrid individuals cannot build their own populations because they have little chance of encountering equal hybrids as sexual partners. Instead, they only encounter the individuals of the two parental species. If, however, the hybrids mate again with the parental species, then this represents a genetic backcrossing. The hybrid genomes blend again with the parental genomes, and except for a limited introgression of a few genes from the foreign species, nothing changes in the two parental species. The two parental species remain preserved, and a third species cannot evolve.

   However, many plants are able to self-pollinate. Thus, they do not need any sexual partners, and therefore, hybrids do not risk mating with the organisms of the parental species, which would eliminate the chance to propagate as a hybrid. Due to self-pollination, however, they are able to build a distinct population that is strong in number, which is reproductively isolated from the parental populations from the start and can compete with these populations because of their own reproduction potency. Accordingly, three species have evolved from two parental species.

   Of course, it must be considered that the new group of exclusively self-fertilizing organisms is not a gene-flow community and thus cannot be a species in this sense. However, many self-fertilizing organisms are not exclusively self-fertilizing all of the time. They occasionally undergo a biparental gene exchange (see above).

   In animals, in contrast to plants, the representatives of few taxa are capable of self-fertilization, for example, many trematodes (flukes) and cestodes (tapeworms). The well-known pork tapeworm *Taenia solium* is almost exclusively self-fertilizing because the human (the final host in the vast majority of cases) can only sustain a single worm. Flukes and tapeworms are therefore candidates for hybridogenic speciation in the animal kingdom. Indications that this has actually happened, however, are rare (Hirai and Agatsuma, 1991).

2) **Tetraploidy**: Self-fertilization, a major difference between plants and animals, is the first important reason why hybridogenic speciation occurs more commonly in plants than in animals. There is, however, a second important reason. If the members of two different species interbreed, then an F1 hybrid results, whose diploid genome consists of the chromosomal sets of two different species. This can lead to disturbances in meiotic chromosome pairing because the chromosome partner available for tetrad formation is from another species, causing misalignment of some chromosomes and deranged chromosome pairing in meiosis. However, a lack of correct tetrad formation means that correct separation

of the genomes during the formation of the haploid germ cells does not occur. Sperm and eggs in the F1 hybrid consequently do not receive complete genetic sets and thus are not fully equipped with all the necessary genes. Alternatively, they receive supernumerary genes or chromosomes, which also may cause sterility. This is one reason why there are postzygotic reproductive barriers between different species. This is also a reason why, in species crossings, the fertility of the species hybrid is affected rather than its vitality. A species hybrid has a higher chance of reaching adulthood and thriving than of actually being fertile (Wu, Johnson, and Palopoli, 1998).

However, it is remarkable that this problem of the disturbed meiotic tetrad formation in species hybrids occurs significantly less often in plants than in animals. There is a particular reason for this. Plants can (for mostly unknown reasons) more easily live in a tetraploid or higher polyploid state than animals. If the plant deviates from the norm and has four or even more chromosome sets in each cell instead of the usual two (diploid), this frequently has no apparent consequences. This is a great contrast from animals. Rough estimates state that one-third of all plant species have a polyploid origin (Schilthuizen, 2001). Many cultivated plants, for example, are tetraploid. If the supplementary chromosome sets stem from the same species, this is called autopolyploidy; if they stem from different species, then this is called allopolyploidy.

In the crossing of two different species, the zygote of the new species hybrid contains two different genomes, which stem from the two parental species: one chromosome set is obtained from the father species, and the other chromosome set is obtained from the mother species. The hybrid is therefore allodiploid. This can, and in fact, usually does lead to disturbances in the meiotic tetrad formation. However, plants can easily become tetraploid and they can bypass the upcoming meiotic disturbances because when every chromosome finds a conspecific homologous partner, then allotetraploid meiotic cells can form entirely normal tetrads. The only difference compared to the meiosis of the purebred parents is that the number of tetrads has doubled in allotetraploid organisms; the tetrads themselves are just like those in the purebred diploid parents.

If such allotetraploid cells in the hybrid organism undergo meiosis and reductional divisions, the resulting gametes are not haploid; instead, they are diploid. Diploidy of the germ cells appears not to block the function of the mature germ cells; thus, allotetraploid hybrid organisms can produce zygotes and vital tetraploid offspring.

Thus, tetraploidy explains why species hybrids in plants are frequently fertile in producing viable germ cells. This still does not explain why hybridogenic speciation is possible in plants. Hybridogenic speciation requires the existence of a barrier against backcrossing with the parental species.

At the same time, tetraploidy also explains this barrier. If a diploid mature germ cell of the hybrid fuses with a haploid germ cell of the parental species through backcrossing, then a triploid zygote is generated, and the offspring of this zygote would be triploid. A triploid organism, however, would then be incapable of meiosis because no appropriate tetrad formation would be possible in triploid organisms. Consequently, the hybrid organisms can only continue to exist if they mate again with equal hybrid individuals. Backcrossing with their parental species would create a dead line of sterile F1 offspring, and they would not be able to continue reproduction.

Now the first peculiarity of plants comes into effect: the ability to self-fertilize. The tetraploid hybrid organisms do not need a sexual partner who may be diploid, a requirement that (in animals) would bring further reproduction to an end. They can reproduce with themselves. The mature allodiploid sperm fertilizes a mature egg of its own mother individual, and therefore this egg is also allodiploid. Thus, the resulting zygote is tetraploid like the mother organism.

Due to a combination of self-fertilization with allotetraploidy, hybridogenic speciation is possible. The capability for allotetraploidy is the second important reason, after self-fertilization, why plants are much more frequently able to generate hybridogenic species than animals. Allopolyploid hybrids are often especially resistant because they combine in themselves the optimal properties of both of their parental species.

Allotetraploid hybrid formation is a remarkable example of a speciation within only a single generation because a post-zygotic barrier immediately comes into being in the first generation of hybridogenesis. This barrier immediately stops gene flow between the hybrid and its parental organisms. In only a single generation, the genome duplication produces plants that can no longer reproduce with their parents, but only with themselves. Furthermore, this is a speciation without any changes in genes. Not a single DNA sequence distinguishes the individuals of the new species from the individuals if its stem species.

6.27
Is the Italian Sparrow (*Passer italiae*) a Hybrid Species?

In Europe, three different forms of house sparrows exist, whose species status is even today still debated: the well-known House Sparrow (*Passer domesticus*), the Spanish Sparrow (*Passer hispaniolensis*) and the Italian Sparrow (*Passer italiae*). The House Sparrow populates all of Europe except for the polar region and, interestingly, Italy and its neighboring islands. The Spanish Sparrow breeds in Spain on the Balkans and on most Mediterranean islands. The Italian Sparrow mainly inhabits Italy, Sicily, Corsica and Sardinia, but remarkably, it is also found on Crete. In most of Europe, the three forms generally exist separately, but there are also broad overlapping regions, such as in north and south Italy, in Spain, in the Balkans and in North Africa. In these overlapping regions, the species coexist side by side without blending in some regions, whereas in other regions, hybridizations occur. In the hybridization regions, there are extended populations of phenotypically intermediate organisms (Töpfer, 2007).

On the Iberian Peninsula, on the Balkans and in parts of North Africa, the Spanish and the house sparrow coexist sympatrically, without any hybridization occurring.

In the region of joint occurrence, they populate separate habitats, with the House Sparrow occupying the traditional habitats, cities and villages and the Spanish Sparrow progressing into rural habitats. In contrast, in Tunisia and eastern Algeria, a hybridization of the two species occurs, with variable populations of sparrows being produced there.

The Italian Sparrow can be found from the Apennine Peninsula to the southern edge of the Alps. A transitory region exists south of the Alps, roughly 35 to 40 kilometers wide, between the populations of the House and the Italian Sparrow, in which hybridizations also frequently occur. In the south of Italy, the Italian Sparrow is connected to the Spanish Sparrow through a broad, smooth transitory region. The animals on Corsica are similar in appearance to those of northern Italy, while the sparrows on Sardinia exhibit distinct transitions between the Italian and the Spanish Sparrow.

This presents a situation that is difficult to interpret. Neither the concept of isolation by distance nor the concept of a ring species can be applied. There is no distinct geographical expansion line along which the continually decreasing genetic compatibility could be traced, as in the case of isolation by distance or in ring species. Moreover, the House and Spanish Sparrows do not occupy separate geographical ranges with only a small contact zone, as in the case of the Carrion and Hooded Crow. Instead, the intermediate regions occur in a ragtag fashion at some locations, and they are absent at others. Third, the phenomenon of partial genetic introgression in an otherwise unambiguous side-by-side coexistence of two species is not realized here, as in the case of the Wolf and the Coyote. Instead, pronounced intermediate populations exist, living in particular geographical regions.

In this book, I have tried to avoid the term "transitional stage" for species status. This term is frequently used for several species with overlapping breeding areas and occasional hybridizations. This practice ignores, however, the fact that a limited introgression of genes among otherwise separate species is a normal process and that in some cases, this even has a biological importance (see below). In several examples, it is not justified to refer to occasional hybridizations and gene flow across the species border as a "transitional stage," indicating the origin of two new species. Open species boundaries may simply be a stable continuous situation in the relationship of two species.

In the rare case of the sparrows, however, we deal with a true case of a "transitional stage," where it is logically impossible to decide whether these are different species. House and Spanish Sparrows are a real borderline case. There are convincing arguments against the species status of house and Spanish Sparrows, even though House and Spanish Sparrows are not unambiguously conspecific.

There is, however, not only the problem of whether House and Spanish Sparrows are species or races. In addition, the Italian Sparrow is often designated as a hybrid species. It is supposed to have originated as a third species from house and Spanish Sparrows. This opinion dates back to Wilhelm Meise half a century ago, who because of the intermediate appearance of the Italian Sparrow, which bears the traits of both the House Sparrow and the Spanish Sparrow, arrived at the conclusion that the Italian Sparrow was a hybrid species (Töpfer, 2007).

However, a hybrid species is something different from a clinal transitory population between two species. A hybrid species is a hybridogenically evolved, new, third species that has emerged from two parental species that had been clearly separated. To speak of a hybrid species, the prerequisite has to be fulfilled that the two parental species have re-encountered each other secondarily after a distinct period of separation and have thereby produced a new species by mutual hybridization. This speciation process absolutely requires that the hybrid species has delimited itself unequivocally from the two parental species by certain properties so that it cannot retroactively blend with them again. The possibilities of backcrossing have to be barred; otherwise, the hybrid species cannot prevail in the long run (see above).

Nonetheless, the Italian Sparrow is connected to House and Spanish Sparrows via clinal transitory populations both in the north and south of Italy. This corresponds to the mechanisms of how geographical races evolve and speaks against the status of a hybrid species. How the Italian Sparrow actually originated is unknown. It could have separated from the neighboring populations of the House Sparrow in the north and the Spanish Sparrow in the south by geographical adaptations to local niches in Italy (Töpfer, 2007).

It is probable that Meise's opinion is wrong, although it appeared so convincing that it remains accepted by many ornithologists more than half a century later. Meise's view has therefore influenced ornithologists considerably. The origin of the Italian Sparrow was even considered a textbook example of speciation by stabilized hybridization. The Italian Sparrow teaches us how effectively a convincingly presented argument can influence the perspective and way of thinking of generations, even if the arguments are not strongly scientifically founded (see also some remarks on Ernst Mayr in this book).

6.28
"Gene theft" between two Species of Galapagos Ground Finches

Now, the following picture of species hybridizations results. Species hybrids are a common occurrence in animals and plants. On the one hand, they are individual occurrences without any evolutionary importance. On the other hand, species hybrids can lead to the extinction of a species. Third, species hybrids also have an important biological meaning. They can lead to the origin of a new, third species, which then combines the properties of both parental species in an advantageous way. This is especially often realized in plants.

There is, however, still a fourth important implication that species hybrids have for evolution. Hybridization is in no way a mostly incorrect sexual contact between different species that is merely tolerated but not supported by selection. As new results of Darwin's finches on Galapagos show, the occasional sexual contact with a foreign species can be necessary for the enrichment of the gene pool of a species.

On the island Daphne Major, two species of Darwin's finches live in coexistence: the larger Medium Ground Finch *Geospiza fortis*, with 200 to 2000 individuals, and the slightly smaller Common Cactus Finch *G. scandens*, with only 100 to 600 individuals (Grant and Grant, 2002). Finch beaks are an example of a phenotype that is especially sensitively controlled by selection (Chapter 5). Even differences in the beak's height of only fractions of a millimeter can provide strong selective advantages or disadvantages against competitors. The finches are then no longer successful in cracking the seeds of a particular size or hardness as nourishment. The beak's height is a genetically controlled trait (Abzhanov *et al.*, 2004).

On the island Daphne Major, periodical fluctuations of the sea current cause a dry period lasting three to five years, which in turn is relieved by an equally long wet period. As a consequence, the vegetation changes rhythmically. In the years with wet weather, the finches are provided with small, softer seeds, and in the dry periods, they are provided predominantly with thick, hard seeds. This phenomenon is accompanied by a remarkable oscillation of beak size in the Medium Ground Finch. In dry years, almost all thin-beaked finches die because their food plants with soft seeds are unavailable. Only the thick-beaked individuals survive because they are able to crack extremely hard seeds (Grant and Grant, 2002). Most of their offspring then have larger and broader beaks, until after a few years, the return of wet weather lets soft-seeded plants grow again. Then, the few surviving thin-beaked finches immediately have a selective advantage, produce higher numbers of offspring and start to become dominant in the total population by replacing the thick-beaked individuals.

This is an example of an allelic polymorphism (Chapter 5). The gene pool of the population of the Medium Ground Finch contains both the alleles for being thick-beaked and the alleles for being thin-beaked. Besides being an interesting example of the fluctuation of allelic frequency distributions, ground finches also show an entirely different and remarkable phenomenon. If, during such a climate change, a population passes through a "bottleneck," it has proven true that in spite of an extensive mating barrier, hybridizations occur to a limited extent between the Medium Ground Finch and the Common Cactus Finch. The hybrids survive well. Because they hardly blend with each other, but predominantly backcross with the parental species, they effectively prevent the two species from merging by blending and thus from becoming separate species. Gene flow between the two species is limited.

However, the occasional interspecies hybridizations appear to be necessary for the survival of the species. We are confronted with the paradox that a process that at first glance is known to endanger the existence of a species is, in this special case, just the contrary: it is apparently necessary for the survival of a species. Interspecies hybridization results in more variable gene pools, which consequently creates a higher genetic flexibility. In this way, the thin-beaked Common Cactus Finch receives alleles from the thick-beaked Medium Ground Finch and thus gains genetic material for the expression of a slightly thicker beak, a genetic resource that may have gotten lost in its own species by passage through the bottleneck. This is a form of "gene theft" from a foreign animal species, by which the fitness of the "thieving" species is strengthened.

6.29
"Gene theft" between two Species of Green Frogs (*Pelophylax ridibunda* and *P. lessonae*)

The well-known European Water Frog (*Pelophylax esculenta*) is a taxonomically problematic case whose status as a species is disputed. The Water Frog is not a distinct species; it is not reproductively isolated from the two species to which it is related, the Marsh Frog (*P. ridibunda*) and the Pool Frog (*P. lessonae*) (Color Plate 8). The Water Frog is diagnostically distinguished from the Marsh Frog and the Pool Frog, but it does not form a separate gene-flow community that is isolated from the other two species.

Pelophylax esculenta results from a hybridization of two species, the Marsh Frog and the Pool Frog. No doubt, the Water Frog is a hybrid. The Water Frog, however, exists only as an F1 product and not as an F2 or even an F3 product. The Water Frog, as an F1 hybrid with the genomes of the Marsh Frog and the Pool Frog, does not continue into a second generation. Instead, the Water Frog must be repeatedly recreated. The hybrid status of the Water Frog initially resembles the many known cases of species hybrids, which result from "accidents" without evolutionary importance, because they do not reproduce among themselves any further but die off again at the end of their individual lives (see above). The Water Frog, however, is not an exceptional transient species. Instead, it is found permanently in most areas.

Water Frogs reproduce very well among themselves, as can be observed in the garden pond. They are vital and completely fertile, but if two Water Frogs reproduce with each other, then only part of the offspring of this reproduction are Water Frogs; other offspring are again pure Marsh or Pool Frogs. The offspring of the Water Frog do not continue a distinct line of Water Frogs. The remarkable hybridization between the Marsh Frog and the Pool Frog does not produce a new line of organisms that would then exist separate from Marsh and Pool Frogs because there is no autonomous Water Frog genome that exists separately from the genomes of the Marsh and Pool Frogs and could pursue a separate evolutionary line. The Water Frog is not a third species alongside Marsh and Pool Frogs that evolved hybridogenically. Instead, the Water Frog evolves again and again anew.

How is this explained? It starts with a hybridization between the two species, Marsh Frog and Pool Frog. The Marsh and Pool Frog mate with each other unrestrictedly; there is no prezygotic mating barrier. In doing so, they do not forfeit their identities. The hybrid, the Water Frog, is vital and fertile. The zygote resulting from the hybridization, as well as all the Water Frog's somatic cells, are allodiploid; they contain a genome of the Marsh Frog and a genome of the Pool Frog. Its germ line cells are allodiploid, but only for a certain time in the ontogenetic development of an individual Water Frog. Before meiosis starts in the testes of the male or in the ovaries of the female, that is to say, before the preliminary germ cells start to differentiate into spermatogonia or oogonia, one of the two genomes is completely removed from the preliminary germ cells. Accordingly, the premeiotic germ cells only contain one of the two parental genomes. Only the somatic cells of the Water Frog are equipped with both genomes. Only the somatic cells are true hybrid cells, but these die when the frog dies.

Consequently, a genetic recombination between the two parental species cannot occur in the germ line during the subsequent meiosis because at the beginning of meiosis, only one of the two genomes is still present. When the Water Frog produces mature germ cells, these sperm or eggs always contain either pure Marsh Frog or pure Pool Frog genomes. Therefore, a male Water Frog ready for mating either produces Marsh Frog or Pool Frog sperm, but no Water Frog sperm, and similarly, the Water Frog female only produces Marsh Frog or Pool Frog eggs. The Water Frog can mate with whomever it wants: (1) a Pool Frog or (2) a Marsh Frog or even with another (3) Water Frog . The offspring are in any case (1) Pool Frogs or (2) Marsh Frogs or again (3) hybrids, that is, Water Frogs.

This is an unusual situation. While the Water Frog is again and again recreated, it is nevertheless not an autonomous new species, as in the hybridogenic speciation of many plants. In the long run, Marsh and Pool Frogs keep their identities, although they continue to intermingle (Schröer and Greven, 1999). The reason for this is that no genetic recombination occurs between the genomes during the hybrid's meiosis. The Marsh and Pool Frog genomes remain preserved unblended.

Water Frogs are effectively "reproductive parasites." To secure their continued existence, the egg of the Water Frog must "steal" the genome from another species in the course of insemination to build the somatic cells of its body. However, this "stolen" foreign genome is not used for the meiotic recombination. The produced offspring does not contain recombined genomes. For this reason, the Water Frog is also termed a "kleptogamic form" or a "kleptospecies" (from Greek *klepto* = to steal) (Dubois and Günther, 1982).

Now, what is the situation regarding the species status of the Marsh Frog *P. ridibunda* and the Pool Frog *P. lessonae*? They cannot be races, for races do not occur permanently syntopically in the same geographical region without intermingling with each other (Chapter 5). What is more important, however, is that no genes flow from the *P. ridibunda* to the *P. lessonae* gene pool via the Water Frog or vice versa. This means that Marsh and Pool Frogs stay clearly separated. Consequently, *P. ridibunda* and *P. lessonae* are each, for good reasons, distinct species.

This example provides another good argument to justify the position that the species concept of a reproductive community is not precise and should be replaced by the more accurate term gene-flow community (see above).

6.30
How many Genes Must Mutate for the Origin of New Species?

It is often assumed that for speciation, the cumulative alteration of many genes is necessary and that only this can lead to the divergence of populations. On the contrary, gene flow barriers can already be created by alterations in only a few genes. Single-gene speciation is possible (Orr, 1991). The two Hawaiian *Drosophila* species

delimitations. However, these barriers are leaky. Many species (even if they belong to different genera) can certainly interbreed. Thus, the genetic cross-compatibility of the members of different species is often not an all-or-nothing matter but a question of quantity (Ghiselin, 1997). This realization is loaded with problems. It means that there are smooth transitions between the species.

Species hybrids are a problem for the definition of the species because hybrids can be assigned to neither one nor the other parental species. At most, a hybrid from two different species belongs to both parental species simultaneously. This fact is used by some authors as an argument against the real existence of species. The open boundaries between gene-flow communities leave room for doubt concerning whether gene-flow communities actually exist as such in reality, as the individual organism is often able to successfully interbreed with a member of different species so that the genes of different gene-flow communities blend into each other to a limited extent. The fact that single individuals belong to neither the one nor the other species is often observed as evidence that species are arbitrarily delimited units, such that these units would not exist beyond human conceptualization.

There are objections against this opinion. Vague boundaries are no argument against the real existence of objects. A cloud exists as a real object, even if it overlaps with a neighboring cloud and many droplets of water can be assigned to neither one nor the other cloud. Colors exist in reality, too. Clouds and colors are not merely mental constructs made by humans, even if there are smooth transitions between clouds or colors. The fact that the boundary between two objects has to be artificially drawn does not necessarily imply that the objects must be artificial constructs.

In many cases, a species can hybridize with another, phylogenetically related species in a particular geographical region, while it cannot do so with the same related species in another region. The quantitative extent of cross-species blending could also vary from region to region. For example, the Pied Flycatcher (*Ficedula hypoleuca*) and the Collared Flycatcher (*F. albicollis*) in Europe show varying extents of crossbreeding in different regions where their breeding grounds overlap. The percentages of mixed matings are significantly different on the Swedish island Gotland than in the Czech Republic (Saetre *et al.*, 1997). Therefore, the genetic compatibility between two species is not a constant property of the species as a whole, but a varying property of the individual subpopulations of this species.

It is not possible to determine a reliable quantitative limit above which the gene flow is so frequent that the organisms of two populations must be classified as belonging to a common species. This intention already fails due to the fact that there are differences in the gene exchange between the sex chromosomal and the autosomal genes. Setting a quantitative limit would mean that in certain borderline cases, the males would all be separate species from each other, while the females in the same region would all have to belong to one and the same species (Dres and Mallet, 2002).

The logical understanding of the species as a gene-flow community can be nicely explained with the following example of two tit species in Europe. The Willow Tit (*Parus montanus*) breeds from France across the entire Palearctic as far as East Asia, presumably without geographic barriers between the populations. Accordingly, all

Willow Tits belong to a gene-flow community where all the organisms are cohesively connected. The Marsh Tit (*P. palustris*) also populates a breeding area that is contiguous from Europe eastwards to the Ural Mountains. However, there is a second breeding area in far East Asia. As these two geographic occurrences are clearly separated from each other, gene flow between European and East Asian Marsh Tits has in all likelihood been disrupted. We therefore have three species: the western Marsh Tit, the eastern Marsh Tit and the Willow Tit.

There is, however, the problem that the Marsh and Willow Tits in Europe hybridize in rare cases (del Hoyo, Elliott, and Christie, 2007). This means that an individual allele in a European Willow Tit has a higher chance of introgressing into a foreign species, the European Marsh Tit, than into the distant members of its own species, the East Asian Willow Tit. Given this, why do all Willow Tits belong to a common species, whereas they are delimited as a separate species from the Marsh Tit?

The logic is as follows. All populations of the western Marsh Tit, the eastern Marsh Tit and the Willow Tit are each continually connected intraspecifically by clinal transitions. The western Marsh Tits apparently form a cohesive gene-flow community, as do the eastern Marsh Tits and the Willow Tits, although the far distant organisms within each of the three species apparently may never encounter each other, may never have sexual contact with each other, and their alleles do not bridge the distance of the entire breeding area. The gene flow connection is given by a continuous gene exchange through several intermediate populations, and this connection between the adjacent "chain links" across the gene-flow community is crucial for group cohesion (Figure 2.7). The concept of the gene-flow community does not imply that the individual alleles ever reach far distant organisms within a species. What counts is only the cohesion as a whole.

7 The Cohesion of Organisms Through Genealogical Lineage (Cladistics)

7.1 Preliminary Remarks on Descent Connection

This chapter deals with the problem of converting phylogenetic trees into taxonomic groups. This is a difficult problem because phylogenetic trees are unbounded continua without borderlines, while taxonomic groups are delimited groups (Mallet, 1995). The cohesion of the individual organisms in a genealogical connection through the parent-child-grandchild lineage is a constantly and steadily progressing continuum without any boundaries. The biological processes that are relevant for the production of an F1 and an F2 (and so on) generation from the parental generation cannot define taxon barriers. Nothing that connects the F1 generation with its parents can define the birth of a species. Genealogical cohesion implies the birth of children and grandchildren, not the birth of taxa. Nothing in the succession of consecutive generations constitutes the end of one taxon and the beginning of another taxon (Simpson, 1961).

Taxonomy means the classification of organisms into groups. Without cohesion, no species can be defined. However, what is just as important and sometimes overlooked is that delimitation is of similar importance. Without delimitation, there can be no species. A species is a group of organisms that are connected to each other and delimited from other such groups. Each group formation requires criteria of cohesion, but at the same time, also criteria of delimitation against neighboring groups (Mishler and Donoghue, 1994).

This reveals the problem of how to delimit species in cladistics. What are the species borders in cladistics? Where does one species begin and another cease? What defines the concept of a speciation in cladistics? Can the bifurcation into two sister branches be defined by the criteria of the genealogical succession of generations, or is species delimitation only possible by the criteria of other species concepts as the typological concept or the gene-flow concept?

At first glance, the cladistic delimitation of species is not perceived as a problem. Cladistics is the theory of phylogenetic bifurcations, and every phylogenetic tree displays the origin of new species as the bifurcation of the evolutionary tree. Horses (*Equus caballus*) and Donkeys (*Equus asinus*) descend from a common ancestor, a

Do Species Exist? Principles of Taxonomic Classification, First Edition. Werner Kunz.

Figure 7.1 Horses and Donkeys descended from a common ancestor, a small equid in the Miocene. This ancestor was a single species, consisting of individuals that all were connected by gene flow. Recent Horses and Donkeys are no longer connected via mutual gene flow. It is for this reason that they are different species, rather than because they are distinguished based on differences in traits.

primitive equid in the Miocene (Figure 7.1). Because horses and Donkeys look different and also look different from the ancient horse, which was indeed much smaller than all currently living *Equidae* (the family of horse-like animals), one could think that speciation is easy to understand. The reasoning would go as follows: The ancient horse has changed and today no longer exists in its former shape, so it has become extinct. Horses and Donkeys differ from the ancient horse and by this reason are considered newly evolved species.

However, such a way of defining delimitations between species is not based on genealogical connections among the individuals. The evaluation of trait changes has nothing to do with descent connection. Trait changes are not a quality of mother-child-grandchild relations. In fact, trait changes are the basis of class formations. Class formation is the sorting of organisms into groups by subjective measures that are based on decisions of the human mind, not on border lines that exist in nature (Chapter 3). Classes are group entities that are ontologically different from relationally connected organisms, which are individuals in the philosophical sense (Ghiselin, 1997). Cladistics, however, in its original purpose, sought to group the organisms by their relational connections, not by their differences in traits.

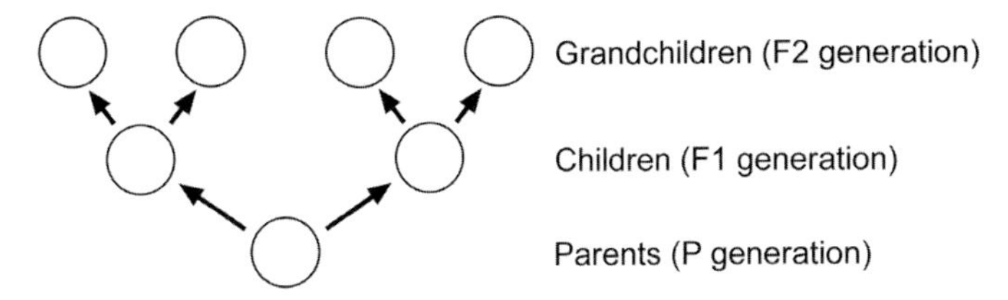

Figure 7.2 The simple phylogenetic (cladistic) tree can only be applied to organisms with uniparental reproduction.

7.2
The Problem of Displaying the Phylogenetic Tree in the Case of Biparental Reproduction

Descent cohesion is usually displayed by a simple phylogenetic tree (Figure 7.2). Such a tree, as has been in use in its form of presentation since Darwin and Haeckel, evokes the impression of a simple biological matter. The ancestors are displayed at the bottom, and the descendants are placed above the ancestors because time flows toward the top of the display. The ancestors are located as a stem at the basis of a bifurcating tree (Figures 2.2 and 2.6). Because reproduction usually goes hand in hand with multiplication, the stem is not just continued as a vertical line toward the top, but the stem bifurcates into sister branches in a V-shaped pattern. This leads to the well-known figure: the phylogenetic tree where the P generation forms a stem that bifurcates into sister branches that, in turn, bifurcate into grandchild branches in the further generation, and this continues until the present time (Figure 7.2).

However, such a tree is something other than a classification into taxonomic groups that should be delimited entities. The question arises of how the phylogenetic tree can be combined with taxonomy. This is not as easy as it sounds at first. How can the phylogenetic tree be transformed into species entities? The phylogenetic tree is a continuum of phylogenetic lineages, whereas species are delimited groups.

The phylogenetic tree consists of two elements: the lineage and the bifurcating fork. The latter is called a "clade" (Figure 2.6). In applying the phylogenetic tree to the species entities, the question arises whether only the bifurcation into two branches can be the *de novo* origin of a species (cladogenesis; see below) (Figure 2.3b) or whether a new species can also originate in the course of an undividedly continuing lineage (anagenesis) (Figure 2.3a).

There are different hierarchical levels of biological organization: the level of genomes, of cells, of organisms and of taxa (see below). Genomes reduplicate into daughter genomes, cells duplicate mitotically into daughter cells, mother organisms produce daughter organisms, and taxa bifurcate into daughter taxa. Any one of these levels has its own phylogenetic tree, which in each case should display reproduction and bifurcation. Genomes bifurcate, cells bifurcate, and organisms and taxa also bifurcate. A phylogenetic tree is always the graphic representation of a replicating and bifurcating system. However, do taxa replicate? Without reduplication and separation, there would not be a phylogenetic tree.

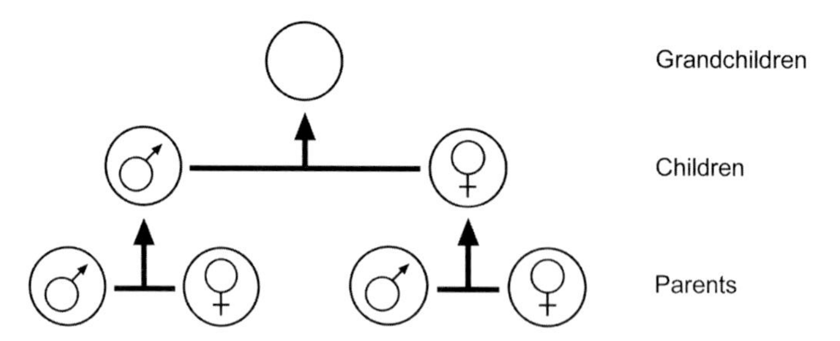

Figure 7.3 Reproduction of biparental organisms cannot be displayed in a simple cladistic tree because the fusion of gametes (fertilization) reverses bifurcations into a network. The branches do not bifurcate in an upward direction along the time scale but, rather, fuse upwards.

Because the phylogenetic tree displays only reduplication and bifurcation, in the first instance, only uniparental reproduction processes can be displayed in a phylogenetic tree. Uniparental reproduction is the reproduction of a single individual without the sexual merging of two parents. It is vegetative reproduction, parthenogenesis or self-fertilization (Chapter 6). Only this kind of simple cleavage of a mother organism into two daughter organisms represents the branching phylogenetic tree as it is displayed in Figure 7.2. Biparental sexuality is not a kind of replication and branching, and thus the reproduction of biparental organisms cannot be displayed in a simple phylogenetic tree (Figure 7.3).

Many single-celled protists, for example, amoebae or flagellates, reproduce uniparentally. Also, the vegetative reproductive stages of sponges and *Coelenterata*, that is, multicellular organisms, as well as the parthenogenetic life cycles of rotifers and water fleas are also uniparental. These replication processes can easily be displayed in a bifurcating phylogenetic tree. Bisexual processes, however, cannot be displayed in a simple phylogenetic tree because sexual reproduction is the opposite of bifurcation. Biparental sexuality is the lateral fusion of separated branches. The phylogenetic tree does not apply to sexually reproducing organisms. The fusion of separated branches turns the phylogenetic tree into a reticular network, which is not a phylogenetic tree, for a phylogenetic tree is expected to be a tree with diverging branches that only diverge but do not reconvene. The diagram in Figure 7.2 cannot be used for the representation of a family's phylogenetic tree. A family's tree (Figure 7.3) includes biparental sexuality, and therefore it cannot be a phylogenetic tree, but rather, it is a reticular network.

Every fusion of gametes (fertilization) reverses the basic element of a phylogenetic tree: bifurcation. The bifurcating phylogenetic tree is tarnished by sexual processes. Biparental sexuality means that the offspring of an individual becomes "contaminated" by its sexual partner. The children of a spouse are only one half its offspring, its grandchildren only one quarter. As shown in Figure 7.3, biparental reproduction reverses the bifurcating phylogenetic tree. The branches do not bifurcate in an upward direction, but to the contrary, they fuse upward along the time scale.

However, there certainly are phylogenetic trees of biparental organisms. How is this possible? This apparent incongruity results from a confusion of the hierarchical level of organisms with the next-higher hierarchical level of species. To understand the application of a phylogenetic tree to organisms with biparental sexuality, one must first realize that it is necessary to switch between different hierarchical levels of biological organization (see below). Because biparentally reproducing organisms do not only cleave into daughter organisms, but also fuse by sexual conjunction, a phylogenetic tree of biparental organisms cannot be displayed at the organismic level; it only can be displayed at the taxon level (de Queiroz, 1999).

To achieve a bifurcating phylogenetic tree for biparentally reproducing organisms, one has to abandon the phylogenetic tree at the organismal level and display the tree at the next-higher hierarchical level. In a biparental phylogenetic tree, the stem and the daughter branches refer to taxa (shaded as a shrouded stem in Figure 6.1), not to organisms. However, the reticular cross-connections within the shrouds refer to branches of the individual organisms; they are not species branches (Figure 6.1).

This consideration makes clear the actual difference between uniparental and biparental propagation (Chapter 6). Biparentality means the transformation of a bifurcating phylogenetic tree into a network. A network means that there are lateral connections, and this is exactly what makes a species. If there are no lateral connections, then the individual organisms are not cohesively linked to each other. If they are not linked, they cannot be species. Only biparentally reproducing organisms can form species in nature. The fusion of phylogenetic branches is almost a definition for that what a species is. Species transform bifurcating trees into reticular networks at the organismic level.

7.3
What are Species Boundaries in Cladistics?

A phylogenetic tree means the arrangement of organisms according to common descent. However, how can a phylogenetic tree define groups? Taxonomy requires the formation of groups. It does not suffice to ascertain that particular organisms belong with each other by sharing a common ancestor. All life on Earth shares a common ancestor. This is not taxonomy.

Group formation presumes not only cohesion criteria but also delimitation criteria. If groups are to be formed, then one also needs rules that specify the criteria on which an entity is delimited from another entity, not only rules that specify their cohesion (Mishler and Donoghue, 1994). By doing so, the problem arises of how cladistics and taxonomy can at all be linked to each other.

The problem of a species concept as a descent community consists of having to find criteria according to which different taxa can actually be defined. The descent cohesion is at first the connection of consecutive generations. Descent cohesion is genealogical cohesion, and genealogical cohesion knows no boundaries. As long as the organisms reproduce, the thread does not break off. Every birth of a daughter organism maintains its current pace without presenting criteria for taxon delimitations.

Descent is the genealogical sequence of parental generations and continuing filial generations. Parents give birth to children, and children give birth to grandchildren.

In no part of the generation sequence do we see a parental generation giving birth to a new species or, especially, to a new genus. If from one mother several daughters descend, then, although a kind of cohesion is given between mother and daughters, this connection does not delimit any taxa. For the mother is in turn also a daughter of her parents, and her daughters give birth again to additional offspring. Accordingly, one can trace back the genealogical cohesion as far as the beginning of all life, without ever having encountered a boundary that could signify a taxon's end.

Because a majority of scientists are convinced that all organisms on Earth have a common root (at least all complex organisms), all organisms are related to each other. Cladistics shows us continuing bifurcations. Yet, what turns a bifurcation into the birth of a taxon? Is the birth of multiple children by the same parents not already a cladistic bifurcation? Where does one species end and the other begin? When does one descent community stop and a new one begin? Taxonomy cannot evade these questions. It is not initially clear how the descent community, which is a progressing continuum, can be linked to taxonomy, which must create delimited groups. Taxonomy faces the difficult task of classifying by common ancestry as well as by group, but a group can only be formed if there are also delimitations against neighboring groups.

Criteria for the decision when one species ceases and a new species begins must be borrowed from other species concepts. Usually species borders are defined through the alteration of traits and/or through the separation of traits (apomorphies, see below). If the traits change, then this is rated as the origin of a new species. If the group splits into two groups with different traits (autapomorphies, see below), then this is rated as the origin of two new species.

Traits by themselves, however, cannot be the reason for classifying organisms into taxa. Traits are never a species definition. They only can be used to distinguish species which previously are defined by other criteria than traits (Chapter 2). A sorting of the organisms according to trait similarity is always a class formation (Chapter 3). Classifying organisms by their traits is something different from classifying organisms by their relational connection, and a descent community is a grouping of the organisms according to relational cohesion. Cladistic taxonomy cannot be typology, the species concept that forms taxonomic groups according to trait similarity.

This difference is elucidated by following example (Figure 7.4). Nine children hold hands with each other and in doing so form a group called *A*. At the end of a certain time (toward the top in the figure), group *A* splits into two separate groups of five and four children (*B* and *C*, respectively). While the group consisting of four children (*C*) remains as it is, the group of five children (*B*) splits again into two additional groups of two and three children (*D* and *E*, respectively). This example makes clear that the children's group cohesion is given only by the fact that they hold hands with each other. To understand what the groups are, which group has originated from which and who is the ancestor of the particular groups, one need not assume that any one of the children changes any of its traits at some point in time.

This example should be understood as a parallel for the descent community in taxonomy. It is supposed to elucidate the fact that the currently living groups

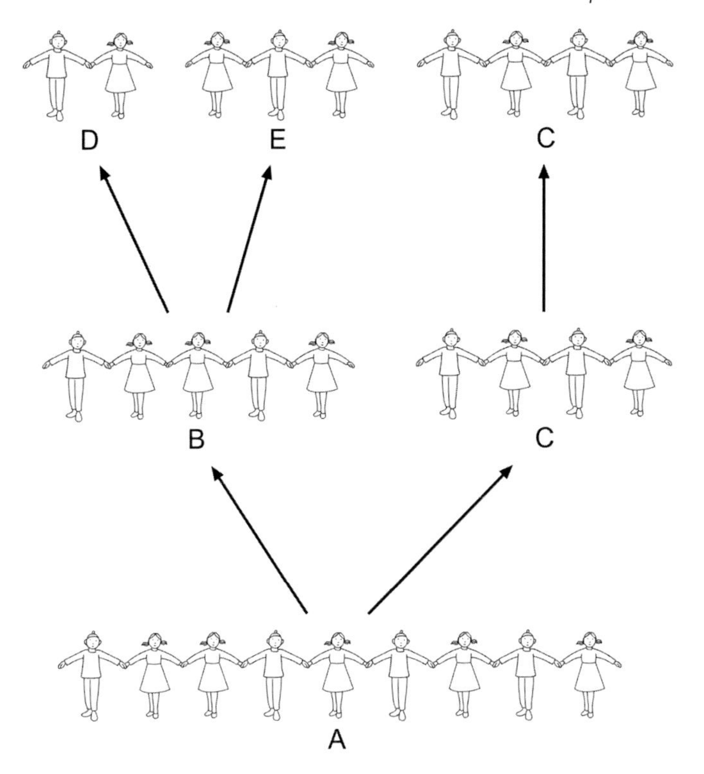

Figure 7.4 Classification of organisms based on relational cohesion in a cladistic tree. Nine children are linked by hand holding with each other and form stem group *A*. Group *A* splits into two separate groups, *B* and *C*. While *C* remains as it is, the group of five children (*B*) splits again into groups *D* and *E*. Group cohesion is provided only because the children are relationally connected. To define cladistic splits, it is not necessary to assume that any of the children has experienced a change in any of his traits. Changes of traits (apomorphies) are secondary evolutionary occurrences that are useful for diagnostic purposes. However, autapomorphies cannot define the phenomenon that is understood to be a cladistic split.

D, E and *C* have a common ancestor, that *D* and *E* have a more recent common ancestor (namely *B*) than *E* and *C* and that *D* is thus more closely related to *E* than to *C*. All these considerations lead to cladistics. The important phrases "split into separate groups," "common ancestor" and "more closely related to each other" can all be understood without a change of traits being necessary in the group members. All groups are defined by the cohesion of their members, not by their traits. Likewise, in cladistics, all branches of the phylogenetic tree are defined by the cohesion of their organisms, not by their traits.

To understand and to define cladistic branching, traits are not needed. There may well exist cladistic branchings without the emergence of any newly derived traits. Introducing trait differences into a cladistic species concept means that the criteria of class formation are mixed with the criteria of the formation of relational groups (see

Section "What is the relevance of differences in traits between two species?" in Chapter 4).

7.4 How is a Cladistic Bifurcation Defined? Apomorphies and Autapomorphies

Earlier in this book, "bifurcations" were repeatedly mentioned. What is a cladistic bifurcation? Descent cohesion only makes statements concerning the organisms' ancestors and offspring along their phylogenetic lineages. As long as the lineage does not become extinct, there is no boundary. A phylogenetic tree by itself cannot comprise taxonomy, as it lacks boundaries.

De Queiroz defines a species in the phylogenetic tree as the segment of a "lineage" between two cladistic bifurcation nodes (de Queiroz, 1999). At first, this sounds simple and convincing, but to understand this definition, a bifurcation node must first be defined. Unfortunately, this is difficult. In the phylogenetic tree, every step of propagation from the parental generation to the daughter generation is a bifurcation node (Figure 7.2). A bifurcation node only delimits the organisms; it does not define a taxon. One bifurcation follows the next, and the clades are repeatedly internested (Mishler, 1999).

To define the split, there are basically two alternatives:

1) Define two daughter branches using the criterion that they can be distinguished by different traits. This practice, however, is only good for diagnostic purposes. It is not suited to define what a species is (see Section "It is one thing to identify a species, but another to define what a species is" in Chapter 2).
2) Define the split into the two daughter branches by the disruption of gene flow between them. This, however, is the species concept of the gene-flow community (Chapter 6). Defining the cladistic split as a disruption of gene flow means that the desired goal of a cladistic species concept, namely, to define the species by the criteria of genealogical cohesion, is not achieved. In fact, the criteria of another species concept are used to define the cladistic species.

1) **The definition of a cladistic bifurcation according to trait differences is as follows**: The first way of defining a cladistic bifurcation consists in the enlistment of trait differences. Species are rated as newly evolved species if they are distinguished by new traits, which the stem species, that is, the ancestor species, does not possess. This is the approach proposed by Hennig, who tried to solve the problem of cladistic bifurcation by introducing the term apomorphy (Hennig, 1966). Apomorphies are newly evolved ("derived") traits that are not observed in the ancestors of this phylogenetic lineage, for example, a color pattern or the form of a beak. Such traits must distinguish the sister branches from the stem group (Figure 2.3c). Otherwise, the cladistic bifurcation would not be a speciation (Figure 2.3b).

   However, the appearance of an apomorphy does not suffice to define a species (Figure 2.3a). The newly derived traits also must distinguish the two sister branches of a phylogenetic bifurcation from each other (Figure 2.3c). In these cases, the new traits are called autapomorphies. Hennig acknowledged only the descendants in the lineages as new species if they were distinguished by autapomorphies. If only an apomorphy newly appears in a phylogenetic lineage, without a bifurcation into sister branches that are recognized by autapomorphies, then this event is not acknowledged as the origin of a new species.

According to Hennig (1966), each phylogenetic bifurcation that is not recognized by the human eye as being characterized by at least two autapomorphies is not considered a split into different taxa (Figure 2.3b). Only if two autapomorphies are recognized (for the human eye) (Figure 2.3c) is this event called cladogenesis, and only cladogenesis is defined to be a split into different taxa (see below). Furthermore, if the organisms of a phylogenetic lineage change their traits, these are, of course, apomorphies because these are newly acquired traits (Figure 2.3a). However, the appearance of an apomorphy without a cladistic branching is also not acknowledged as origin of a new taxon. This event is called anagenesis (see below), and anagenesis is not defined as origin of a new taxon (Hennig 1966).

Altogether, this is a rather subjective proceeding. The choice whether a trait is an apomorphy depends on human decision. In principle, all the descendants of two parents show "newly evolved" (derived) traits. Of course, a mother's two daughters distinguish themselves by different traits; otherwise, the daughters would not be individuals, that is, unique, singular beings. Why are these different traits not all autapomorphies? And why are the two daughters not new species? Each choice of an apomorphy selects one trait out of many available traits, and this choice is subjective because there are no rules for when traits define a species and when they do not (Chapter 4). It is predominantly due to pragmatic considerations that some newly evolved traits are declared to be autapomorphies while others are not.

Furthermore, the occurrence of intraspecific polymorphism is nothing other than the appearance of newly evolved derived traits, that is, apomorphies. These clearly are apomorphies and even autapomorphies that have nothing to do with the origin of new species, because intraspecific polymorphisms are different morphs within a species (Chapter 5). This means that there exist autapomorphies which define new taxa, but there also exist autapomorphies which do not define new taxa. What is the difference of both kinds of autapomorphies?

What are apomorphies? First, an autapomorphic trait is defined as a trait that defines the split into two different daughter branches. However, the split into two different daughter branches is defined as a bifurcation of the evolutionary tree that is defined by the appearance of new autapomorphies. Hence, the whole matter entails circular reasoning.

Furthermore, the use of apomorphies as a criterion of speciation confuses the epistemic approach of a species diagnosis with the ontological approach of the definition of that which a species is (see Section "It is one thing to identify a species, but another to define what a species is" in Chapter 2). Apomorphies are only tools for the diagnosis of species, but they cannot define what a species is.

The concept of "*differentia*" should not be confused with the concept of "*definitio*." Diagnosis is not definition. Only if their species status has been established by other criteria in advance can apomorphies be used to identify two sister branches in a cladogram. If their species status is not known already, apomorphies can never be used to define two sister branches in a cladogram as being two different species.

Recently, the term "specifier" has been used instead of Hennig's "apomorphy" (Mishler and Brandon, 1987; de Queiroz and Gauthier, 1990; de Queiroz, 1998). It could hardly be more obvious that an additional, literally "species-making" label is introduced here. Only by this means can cladistics be turned into taxonomy. With the introduction of the apomorphies or the specifiers, the origin of new species is no longer discovered, but new species are constructed by the human mind. A subjective component has entered cladistics, as it depends on human taste which traits are acknowledged as apomorphies and which are not. Traits must be found that have a certain "validity" to be suited for use as a species definition, but there is no objective criterion for validity judgments such as these in determining why a trait is appropriate to be an apomorphy and why it is not.

The decision of defining two daughter species as newly derived species by selecting distinctive traits is a fall-back into typology (Atran, 1999). It is the reintroduction of the principle of subjective evaluations through the back door into the concept of cladistics, a concept that initially distances itself from this kind of subjective assessment. In principle, Hennig did not solve the problem of the cladistic split, which is the bifurcation of a group of common descendants into two separate groups.

2) **The definition of a cladistic bifurcation by means of the gene-flow community is as follows**: A cladistic bifurcation can also be defined without apomorphies, namely, if the stem group separates into two daughter groups upon disruption of the gene flow. If a species is defined as a gene-flow community between organisms, then the cessation of this cohesion suffices as a criterion of defining when a species ends and when two new ones begin. In biparental organisms, a species then ceases to exist if gene flow is disrupted.

However, this definition is borrowed from another species concept. Defining the end of a species and the origin of two new species by the interruption of gene flow is not a definition through the criteria of the genealogical descent cohesion. Therefore, there cannot be a "cladistic species."

7.5
Descent is not the Same Thing as Kinship: The Concepts of Monophyly and Paraphyly

The term "descent" is not the same as the term "kinship." At first glance, both terms appear to mean the same thing because individuals of a common descent are also related to each other. Life on Earth has only one common descent stretching back through the history, but life on Earth can be subdivided in groups of different degree of kinship. The concept of descent cannot be gradated, but the concept of kinship can.

While the statement "several organisms have the same ancestor" should not be understood quantitatively (there is no stronger or weaker descent), the statement "several organisms are related to each other" should certainly be understood quantitatively because there are closer and more distant kinships.

This becomes clear in the following simple example. All son/daughter descendants (F1 descendants) of a pair of parents are equally related to each other. The four grandchildren of a pair of parents (F2 generation), however, are no longer equally related to each other (Figure 7.2). They are either siblings or cousins and so are related to each other to different degrees. While the four grandchildren in Figure 7.2 form a coherent descent community, they separate into two groups as kinship communities: siblings and cousins. If a number of organisms were grouped according to equal-ranking descent, then this would result in a different group than if a number of organisms were grouped according to equal-ranking kinship. The group of organisms possessing equal-ranking descent only in the F1 generation (son/daughter generation) is congruent with the group of organisms possessing equal-ranking kinship, but it is no longer congruent in the F2 generation (grandchild generation).

This line of thought is the starting point for understanding the concept of the monophylum, which contains more than just a group of organisms with a common descent. Besides having a common ancestor, the monophylum must be complete (Figure 7.5a). A monophylum is a descent community that has to fulfill two conditions. First, it must be a group of descendants with a common ancestor. Second, all the descendants of this common ancestor must belong to this group without exception, for there is no conclusive justification for excluding some portion of the descendants from the group and to include them in another group. For this reason, Ereshefsky stresses: "A [monophylum] must contain a single ancestral species *as well as* all and only its descendant species" (Ereshefsky, 1999). A monophylum always has to follow the rule that all sister taxa must be of the same rank.

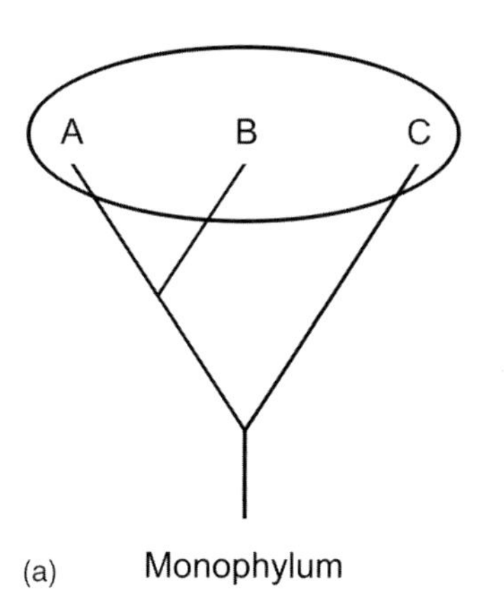

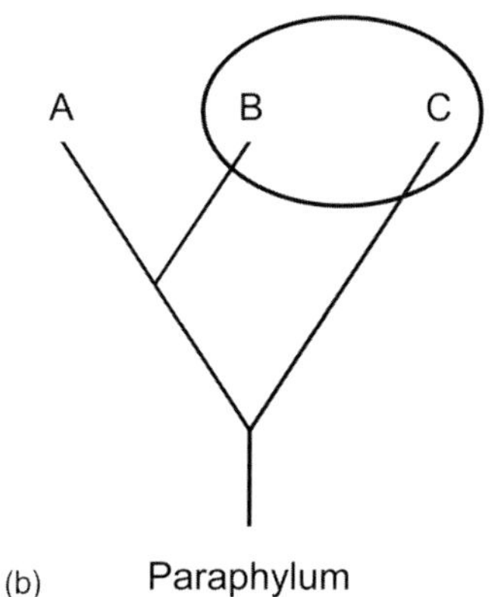

Figure 7.5 A monophylum is a group of descendants with a recent common ancestor. In addition to having a common ancestor, a monophylum also must contain all of the descendants of the common ancestor. If a branch is excluded, the remaining group is a paraphylum. In a paraphylum, the sister taxa (*A* and *B* + *C*) are of unequal ranks.

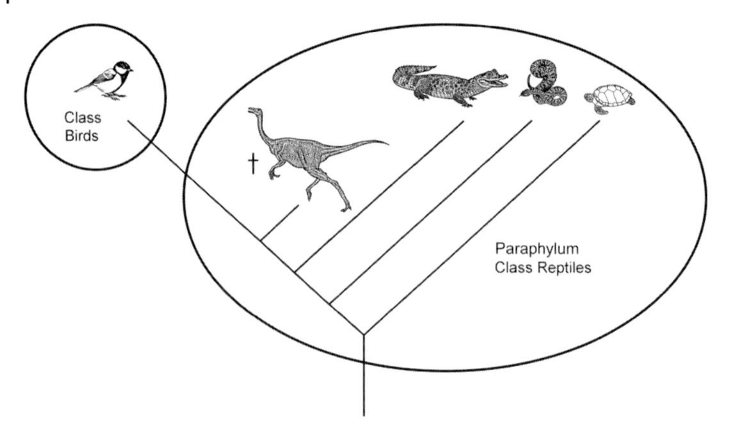

Figure 7.6 If birds are excluded from the class of reptiles, the remaining reptiles are a paraphylum. A paraphylum is a truncated taxon because it is incomplete. Paraphyla are the result of mixing different classification principles, and therefore they are inconsistent and contradictory. The attempt to form a group according to kinship is mixed with the other attempt to form a group according to trait similarity, which is based on subjective standards.

By excluding certain descendants, the rest of the group would no longer be a monophylum, but a paraphylum (Figure 7.5b). The taxonomical principle that permits this kind of group formation is called paraphyly. Paraphyly conforms to the principle of common descent, but it ignores equal kinship in favor of other intentions that are highly questionable. Paraphyletic group formations are possible on many taxonomical hierarchy levels. It is possible to form paraphyla on the level of the taxonomic class, the family, the order or the genus. For example, the reptiles are a paraphylum on the taxon level of the class because the birds are excluded from the class of reptiles (Figure 7.6). The great apes (*Pongidae*) are a paraphylum on the level of the family because they exclude the family of humans (*Hominidae*).

Likewise, the concepts of mono- and paraphyly can be applied to more than just taxon trees; they can basically be applied to all phylogenetic trees. Paraphyletic groups also exist on the hierarchy level of organisms, on the level of cell genealogy and on the level of the replication of DNA molecules (see below). However, it is not possible to apply the term paraphylum to a species, as long as the species is understood as a gene-flow community. A paraphylum must consist of bifurcating branches, and a gene-flow community does not consist of branches (see below).

A monophylum is always a complete combination of all the descendants of a most recent common ancestor. A monophylum always must follow the rule that sister taxa must be of the same rank (Hennig, 1966). The easiest way to understand the monophyletic grouping approach is by the following procedure. From a branching phylogenetic tree, choose some end points of the top horizontal row (e.g., *B* and *C* in Figure 7.5b). Then, go downward in the tree and look for the most recent common ancestor of these end points. Then, the monophyletic group to be formed must contain without exception (when going back upwards) all the descendants of this

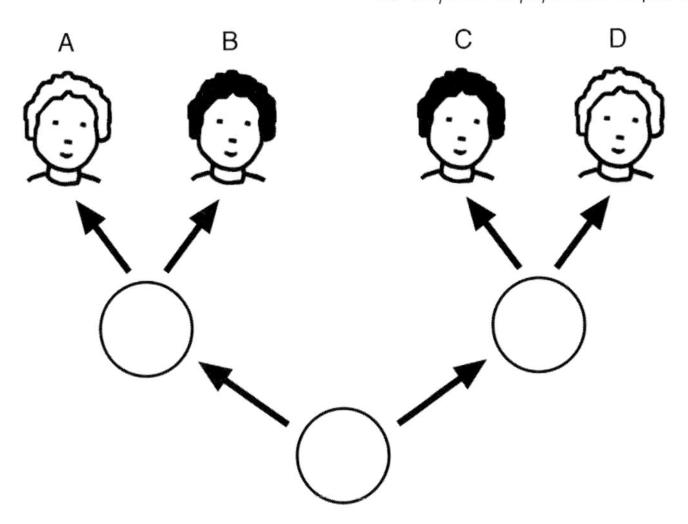

Figure 7.7 Paraphyletic classification ignores the principle of equal kinship in favor of trait similarity. If, in this example, $A + D$ and $B + C$ would be classified into two separate taxa, because of similarity in hair color, this would mean that brothers are ripped apart, but cousins are joined in common taxa because hair color is overrated. Nobody would group the organisms in a family tree in this way, but this is what happens in paraphyletic taxonomy.

most recent common ancestor. Only then is the formed group a monophylum. This is not the case in the example $B + C$ in Figure 7.5b. In this case, the group $B + C$ accordingly is a paraphylum because a lateral group (A) has been removed from the taxon and made into a separate taxon. The monophylum comprises all descendants, that is, A, B and C (Figure 7.5a).

Both groups, the monophylum and the paraphylum, are groups with common ancestors. The difference is that the monophylum is complete, while the paraphylum is missing some part of the common descent.

7.6 Why are Paraphyla used Despite their Inconsistency?

In any case, paraphyly is an inconsistent taxonomical classification principle. Paraphyly disregards the principle of equal-ranking kinship. To withdraw a branch from a monophyletic clade ignores the important principle that sister taxa always must be of the same rank. Paraphyly means that brothers are ripped apart but cousins joined (Figure 7.7).

A paraphyletic grouping has to be refused by everyone who desires consistent grouping that is free of contradictions. It is incomprehensible why a descent community should be displayed incompletely in a taxon. A taxon of reptiles, a taxon of great apes or a taxon of ferns (*Pteridophyta*) should not used in taxonomy. Why, then, are paraphyla used at all despite their inconsistency? There must be some reasons why paraphyly cannot be eradicated. What justification is there for a paraphyletic taxonomy? There are only pragmatic reasons for this; there is no consistent underlying theory.

Only the human desire to combine organisms into a common group that look similar to the human eye can motivate the use of paraphyletic groups (Figure 2.4: $B + C$ is a paraphyletic group). This is not at all a scientific line of thinking. The formation of paraphyletic groups means that two different classification principles are intermixed: grouping according to the descent relation and classification according to trait resemblance.

The best-known example of paraphyly in biology is the aforementioned class of reptiles along with the class of birds within the phylum of vertebrates (Figure 7.6). The reptiles are an incomplete, truncated taxon because the birds that branch off from the rest of the reptiles at an evolutionarily later time have been withdrawn from the reptile class. However, the birds are the sisters of the dinosaurs and are closely related to these, while both together (the birds and the dinosaurs) are less closely related to the crocodiles, snakes and turtles than birds and dinosaurs are to each other.

Thus, the formation of the class of reptiles without the birds contradicts the logic of a kinship-oriented taxonomy. This is not taxonomy according to kinship, but taxonomy according to the assessment of traits. However, the assessment of traits is based on subjective standards that are set by humans ourselves. This means that paraphyla (in contrast to monophyla) are artificial groups, whose coherence and boundaries as separate groups are not demonstrated in nature (Figure 2.4: $B + C$ is a paraphyletic group). For the empirically driven scientist who wants to restrict himself to phenomena that can be objectively observed in nature, paraphyly is a group formation that is highly questionable, if not to be rejected altogether.

We humans think that birds simply look very different from the reptiles, but we are unable to quantify this distinction. Birds are warm-blooded, have feathers, are usually multicolored and often sing in a manner pleasant to the human ear. Accordingly, in the eyes of humans, they distinguish themselves drastically from crocodiles, snakes and turtles. These are cold-blooded, have scales or horny scutes, are usually colored inconspicuously, and make utterances that are almost without exception unpleasant to humans. The qualities of these criteria cannot be reason to group birds and reptiles into separate equal-ranked taxa.

For the modern, factually driven scientist, it is disconcerting that these kinds of criteria serve as a reason to separate the birds from the reptiles and award them an equal rank along with the rest of the reptiles. Why do the possession of a plumage and warm-bloodedness rank so high that the birds are ascribed such dissimilarity to the remaining reptiles? By reverting less to subjective human perception and instead perhaps valuing certain metabolic enzymes that are not visible to the human eye, the birds would not be such a unique case compared to the rest of the reptiles. Then, the birds' autonomy would be ended, and the birds would unambiguously belong to the reptiles, with which they are closely related and from which they should not be segregated.

Let the contrariness of a trait-based and a kinship-based grouping once again be demonstrated by the following example (Figure 7.7). Of the four grandchildren

A through *D*, the two grandchildren *A* and *D* both possess the common trait of having blond hair, while *B* and *C* are dark-haired. According to the principle of close kinship, the four grandchildren must be assigned to the two groups $A + B$ and $C + D$ (the respective brothers). Applying, by contrast, the principle of trait similarity, however, the four grandchildren must be assigned to the two groups $A + D$ (the blond ones) and $B + C$ (the dark-haired ones). Nobody would group the organisms in a family tree in this way, but paraphyletic taxonomy is conducted in this manner.

The principle of a consistent monophyletic grouping was an inherent part of the fundamental theory of the cladistic system of the German zoologist Willi Hennig (1966). The charm of the cladistic system is its clear and consistent logic. Hennig meant to keep subjective criteria out of biological systematics and to model taxonomy according to criteria that correspond to nomological status and not human wishes for feasibility. Hennig meant to specify a grouping that is based only on descent and kinship and is logically conclusive. This outstanding merit of Hennig appears to be underestimated till recent days.

It is not easy to understand why Ernst Mayr, in contrast, so often defended paraphyly (Mayr and Ashlock, 1991) because Mayr, on the other hand, was a decisive advocate of objective science against subjective typology. In retrospect, it is unknowable why Mayr justified the principles of a paraphyletic classification. Because Mayr refused typology in systematics, it is mysterious why he was a firm advocate of paraphyletic systematics. The retention of reptiles as a distinct class was justified by Mayr by claiming that reptiles look entirely different from birds and, therefore, are intuitively perceived as an enclosed group and that birds are immediately conceived as a group separate from this one, even by laypeople in the field of taxonomy (Mayr and Ashlock, 1991).

Mayr defended paraphyly and distinguished it energetically from pure cladistics. He even went as far as to accuse Hennig of having a one-sided perspective because Hennig would ignore the relevance of trait alteration in evolution. Mayr lamented that Hennig's principle of monophyly had caused "painful upheavals" in taxonomy (Mayr, 1982). This is doubtless true, but unavoidable, if the consistence of reasoning is to be given priority over convenience and pragmatism.

Considering both trait differences and kinship relationships at the same time, Mayr rejected the three-kingdom system of living beings introduced by Carl Woese, in which all living beings are classified into bacteria (older name: eubacteria), archaea (older name: archaebacteria) and eukaryotes (Woese, 1990). Mayr – in contrast – defended the two-kingdom system, that is, the classification of all organisms into only prokaryotes and eukaryotes, regardless of the nearly nonexistent kinship of the bacteria with the archaea (Mayr, 1998). The only reason to classify bacteria and archaea into a common group is their phenotypically similar appearance, as both types of organisms are extremely small, have no cell nucleus and do not undergo mitosis. However, these criteria are subjective human estimates that are not scientifically founded.

Mayr accused the cladisticist Willi Hennig of turning classification into an "intellectual exercise" instead of a means of rediscovering information (Mayr, 1982). This was meant derogatively. This opinion did not appreciate the intellectual capacity of Hennig's theory. In taxonomy, the incompatibility of a purely pragmatic goal with consistent theoretical thinking cannot be expressed more clearly. The author of this book has the opposite opinion. He sees a greater degree of scientific merit in mental consistency.

However, the classification of organisms according to consistent monophyletic principles would clearly lead to a system that is significantly more difficult to manage than the current system, and the system of the kingdom of animals and plants would have to be completely rearranged in a new way. This would be highly inconvenient. However, in taxonomy logical consistency and manageability are incompatible (Hull, 1997), which is the reason that two different taxonomies are probably needed (Chapter 2).

The mixture of trait-oriented classification principles with cladistic classification principles in many individual cases leads to the designation of biological taxa that, although comfortably manageable, cannot be theoretically justified. In such a case, it does not help to appeal to the claim that evolution consists of both trait alterations and bifurcations as a justification (Mayr, 1982). The blending of trait-oriented classification principles with cladistic classification principles leads to artificial groupings that reflect nothing other than human convenience. This cannot be called science. A taxonomy based on the assessment of traits, as defended by Mayr to justify paraphyly (Mayr and Ashlock, 1991), cannot be false in principle; therefore, it is not falsifiable and accordingly is not a scientific proposition (Hull and Ruse, 2007).

7.7 Monophyly and Paraphyly on Different Hierarchical Levels

The principle of monophyly or paraphyly can be applied to phylogenetic trees at all hierarchical levels of biological organization, including taxon trees, organismal trees, cell trees or genome trees. A taxon group, an organism group, a cell group or an allele group can be monophyletic or paraphyletic. However, the different hierarchical levels should never be mixed with each other or combined into a common system. Researchers should always explicitly clarified what is meant if a phylogenetic tree is presented. Phylogenetic correlations at the DNA level (gene trees) do not necessarily allow reliable conclusions to de drawn regarding the same phylogenetic correlations at the taxon level (species trees).

The problem lies in the fact that phylogenetic trees at different hierarchical levels are not congruent with each other. Each of the different hierarchical levels, whether taxa, organisms, cells or genomes, has its own phylogenetic tree. It can be faulty to make an inference from the phylogenetic tree of a gene sequence regarding the phylogenetic tree of the respective species. Likewise, it can be faulty to make an inference from the phylogenetic tree of a cell genealogy with respect to the phylogenetic tree of the organisms emerging from it (Lee and Skinner, 2008).

There is a simple reason for this situation, which is demonstrated in Figure 7.8. From the mother object *a*, the two daughter objects *b* and *c* descend. From *b* and *c*, the F2 descendants *d, e, f* and g arise. This scheme of a descent pattern can be applied to

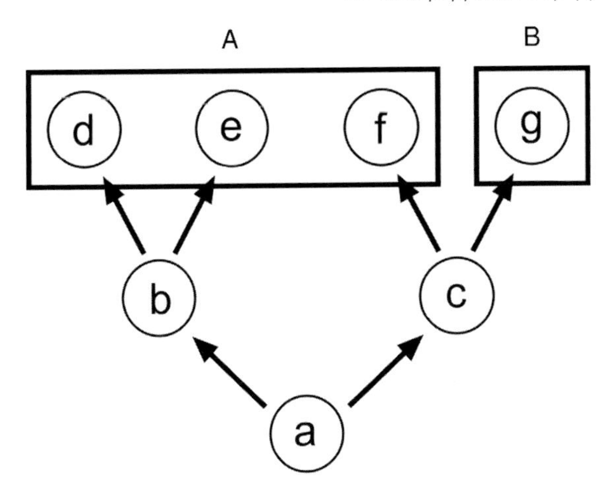

Figure 7.8 Phylogenetic trees at different hierarchical levels are not congruent with each other. A mother cell (*a*) produces the four F2 descendants, *d* through g. At the next highest hierarchical level (the level of the whole organisms *A* and *B*), the cell group *d* through *f* is integrated into daughter organism *A*, while cell g is integrated into daughter organism *B*. Therefore, daughter organism *A* is a paraphylum with respect to the genealogy of its cells. This example shows that the attempt to transfer a phylogenetic tree from one biological hierarchical level to another hierarchical level leads to uninformative results.

genes, cells, entire organisms or taxa. Let us assume that the objects are cells that cleave mitotically. A mother cell, *a*, produces the four F2 descendants *d* through g. At the next higher hierarchical level (the level of whole organisms), the cell group *d* through g is divided, resulting in two separate daughter organisms, *A* and *B*. *A* and *B* are two multicellular daughter organisms that arose via vegetative reproduction through simple cleavage of the mother organism. The first daughter organism, *A*, emerges from the cell group *d* through *f*, whereas the other daughter organism, *B*, emerges from cell g.

From this scheme, it follows that the cleavage of a mother organism into two daughter organisms does not follow the phylogenetic tree at the hierarchical level of its cells. The daughter organism *A*, which originates from the cells *d*, *e* and *f* in the diagram in Figure 7.8, is a paraphylum with respect to the genealogy of its cells. Only a subset of the cellular descendants of the most recent common ancestor cell, which is mother cell *a*, give rise to daughter organism *A*, whereas a more recent side branch, also descendant of mother cell *a* (cell g) produces daughter organism *B*. This has the consequence that the closest relative of cell *f* (belonging to organism *A*) is cell g, which belongs to a different organism (*B*), whereas cell group *d* and *e*, together with cell *f*, belong to the same organism, *A*, although they are much less closely related. Accordingly, organism *A* is a paraphylum at the level of the genealogy of its cells, just as reptiles are a paraphylum at the hierarchical level of the genealogy of classes.

Imagine simple asexual budding of a daughter organism from a freshwater *Hydra* polyp (Figure 7.9). As in the scenario shown in Figure 7.8, one of the two organisms (in Figure 7.9 the right one) consists of cells that are clearly a paraphylum.

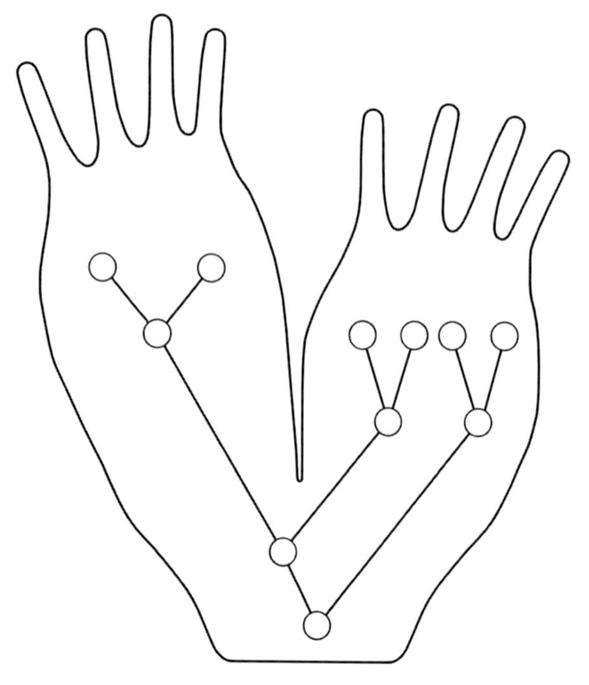

Figure 7.9 Vegetative budding of a daughter organism from a freshwater *Hydra* polyp. The daughter organism consists of cells that clearly represent a paraphylum at the level of the cell genealogy.

These considerations show that the attempt to transfer a phylogenetic tree from one biological hierarchical level to another hierarchical level leads to uninformative results. What is observed as a paraphylum at the level of the cell genealogy does not make any sense if it is transferred to the level of the phylogenetic tree of organisms. Whole multicellular organisms do not necessarily need to have the same phylogenetic tree as the cells of which they are composed. The concept of mono- or paraphyly should always only be applied to genealogical trees within a given hierarchical level (Lee and Skinner, 2008; de Queiroz, 1999), that is, either only to the level of cell trees or the level of organismal trees. To apply phylogenetic trees from one hierarchical level to a different level does not lead to informative insights.

7.8
Gene Trees are not Species Trees

For the same reason, gene trees are not species trees. It can be misleading to make an inference from the phylogenetic tree of a gene sequence regarding the phylogenetic tree of a species, although this is often done in publications addressing gene sequence alignments. However, the hierarchies cannot be mixed.

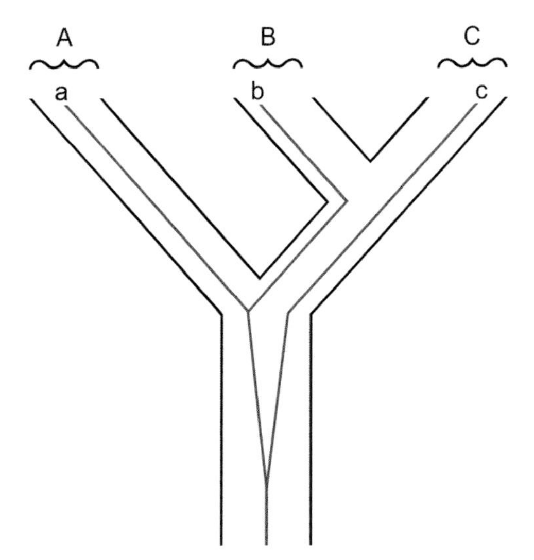

Figure 7.10 Gene trees are not species trees. The timing of a split into species (*A, B, C*) may be evolutionarily younger than a pre-existing allelic polymorphism (*a, b, c*). Therefore, the age of the gene tree bifurcation does not allow explicit conclusions to be made about the age of the species bifurcation in each case.

Species almost never descend from a single pair of parents (Adam and Eve). Instead, each species descends from a founder population consisting of multiple organisms. This stem population may contain a preexisting allelic polymorphism that enters into both daughter species during the split of the daughter species. This split into species may be evolutionarily younger than the preexisting allelic polymorphism (Figure 7.10). Accordingly, the age of the gene tree bifurcation does not allow explicit conclusions to be made regarding the age of the species bifurcation in each case because the gene tree bifurcation may have occurred earlier.

The multiple alleles of a gene may have varying degrees of kinship. If a stem species bifurcates into daughter species (*A, B* and *C* in Figure 7.10), the alleles that enter the daughter species (*a, b* and *c* in Figure 7.10) may already have bifurcated at earlier evolutionary time points, and it may be completely misleading to conclude the age of the species bifurcation from the age of the allele bifurcation.

If the DNA sequence of allele *a* in species *A* is aligned with the sequence of allele *b* in species *B* (Figure 7.10), a close similarity in the DNA sequences is observed because alleles *a* and *b* diverged from each other only a short time ago. If, however, the sequence of allele *b* in species *B* is aligned with the sequence of allele *c* in species *C*, a greater difference in the DNA sequences is observed because alleles *b* and *c* diverged earlier than *a* and *b*.

A taxonomist who examines the degrees of kinship among species *A, B* and *C* could arrive at incorrect conclusions if he only compared one or a few alleles with regard to their sequence similarity. He could conclude that species *A* and *B* are more closely related to each other than to C because the examined alleles only exhibit this kinship relationship. However, the actual case is exactly the opposite: species *B* and *C* are more closely related to each other.

This example shows that hierarchies may not be mixed with each other. Gene trees are not species trees (Nichols, 2001). It is not possible in all cases to infer the phylogenetic ages of organisms whose genomes contain two orthologous alleles from the phylogenetic ages of these orthologous alleles. Gene trees and species trees should not be intermingled into a common phylogenetic system. These types of trees represent different hierarchical levels that are linked to different histories and should therefore consistently be kept separate with regard to datasets.

The discrepancies between gene trees and species trees only become smaller when evolutionarily old species are compared. This is because allelic polymorphism disappears in the course of longer evolutionary periods of time, as most mutant alleles are lost due to selection or genetic drift (Chapter 5). As a consequence, the discrepancies between gene trees and species trees diminish. Eventually, a state is reached at which the phylogenetic tree of alleles matches the phylogenetic tree of the species.

However, not all allelic polymorphisms are evolutionarily short lived. Stable allelic polymorphisms are exceptions to this pattern (Chapter 5). For example, the major histocompatibility complex (MHC) of higher vertebrates consists of a group of genes that all exhibit high allelic diversity. This allelic diversity predates the origins of species. Although humans and chimpanzees split into separate species only approximately five million years ago, the origin of the many variations in their MHC alleles goes back further (Figueroa, Günther, and Klein, 1988). The MHC allelic tree does not reflect the species tree of most higher vertebrates.

Faulty inferences from gene trees to species trees can also be lessened if comparison of DNA sequences is extended to several genes. However, as emphasized previously in other passages in this book, it does not matter that a procedure leads to correct results in most cases. This book is about the foundations on which the theory of taxonomic classification is based, regardless of the fact that non-observance of such foundations leads to the appropriate result in most cases.

The problem regarding the discrepancy between gene trees and species trees leads back again to the problem with paraphyletic structures that occurs if different hierarchical levels are mixed. With respect to the hierarchical level of species, species *B* and *C* in Figure 7.10 clearly form a monophylum. However, their alleles *b* and *c* represent a paraphylum because allele *a* has been excluded from the monophyletic group of alleles $a + b + c$, although it shares a most recent common ancestor with *b* and *c*.

7.9 The Concepts of Monophyly and Paraphyly cannot be Applied to Species

Dissension reigns regarding whether the cladistic concept of a mono- or paraphylum can be applied to the lowest category of taxonomy: species. Hennig (1966), the founder of cladistic taxonomy, avoided applying the cladistic unit of a monophylum to the category of species and restricted the application of the monophyly concept to higher taxa (Hull, 1997; Ereshefsky, 1999). In contrast, Hennig's successors have extended the concept of monophyly to the level of species (Ax, 1995; de Queiroz, 1999; de Queiroz and Donoghue 1988; Mishler and Donoghue, 1994).

The problem in this case is that the concepts of monophyly and paraphyly are tied to bifurcations (Figure 7.5). Without bifurcations, there is no mono- and no paraphyly. However, biparental species as a gene-flow community do not represent a bifurcating system. Within a unit in which reproductive connections between organisms exist, there can be no bifurcations. A bifurcation within a species always splits the species into two separate gene flow communities, thus representing the origin of two new species.

Accordingly, the terms monophyly and paraphyly cannot be applied to biparental species because there is no bifurcation within such species. Mono- or paraphyla must be groups that include distinct branches. The logic of the paraphylum always requires at least two bifurcations, which lead to three branches with dissimilar kinship relationships that would then be expressed in the paraphylum (Figure 7.5). Two branches alone cannot form a paraphylum; and a gene-flow community that always is a group of organisms without any separated branches, never can be a mono- or paraphylum. Bifurcations only exist at higher levels of taxonomic hierarchies above the species level.

This is explained by the following example (Figure 7.11). On the Canary Islands of Tenerife and Gran Canaria, there is an endemic species of finch that only occurs on these two islands, the Blue Chaffinch (*Fringilla teydea*). The Blue Chaffinch is a sister species of the Common Chaffinch (*Fringilla coelebs*). It is thought that the Blue Chaffinch separated from the continental Common Chaffinch only a relatively short time ago evolutionarily.

Let us assume that the stem species of the contemporary Common Chaffinch and Blue Chaffinch inhabited a wide geographical range from northwest Africa through Europe to West Asia, similar to the distribution of the contemporary Common Chaffinch. Let us also assume that during that time, different races arose that remained connected to each other via continuing gene flow (open circles within the stem species in Figure 7.11). When population *e* then reached the Canary Islands, the gene flow cohesion with the continental populations ceased due to allopatry.

Two sister species arose: the recent Common Chaffinch and the Blue Chaffinch. It does not make sense to view the Common Chaffinch as a paraphylum, as its current races, *a* through *d*, are connected by gene flow and, thus, belong to a single gene-flow community. Gene flow cohesion excludes the possibility of a more recent common ancestor or a less recent common ancestor of populations *a* through *e* because all populations are connected and do not consist of a system of separated branches. Therefore, the Common Chaffinch cannot be a paraphyletic species. The concepts of monophyly and paraphyly cannot be applied to species as long as species are understood as gene-flow communities.

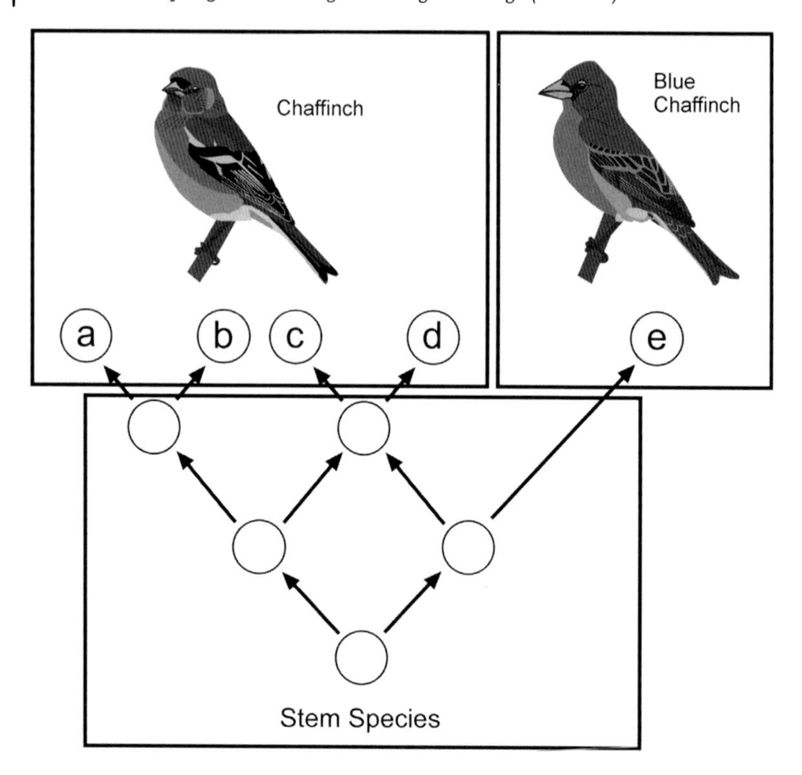

Figure 7.11 The terms monophylic or paraphylic cannot be applied to species. The Blue Chaffinch (*Fringilla teydea*) of the Canary Islands is a sister species of the Common Chaffinch (*Fringilla coelebs*). The Blue Chaffinch (*e*) presumably separated from the Common Chaffinch (*a* through *d*) recently at a time when the stem species was distributed over a wide geographical range through Europe and West Asia and probably subdivided into several races. However despite the fact that the contemporary Common Chaffinch consists of a number of currently living races (*a* through *d*), it is not possible to consider the Common Chaffinch as a paraphyletic species, as all of its races are connected via gene flow and, thus, belong to a single gene flow community. The concepts of monophyly and paraphyly are tied to bifurcations, and a gene flow community cannot be comprised of bifurcations.

7.10 Paraphyly and Anagenesis are Mixed Classifications

"Mixed classification" means that objects are grouped based on the simultaneous application of different classification criteria. Different grouping principles are applied for a given grouping process. In taxonomy, mixed classification indicates that different species concepts based on different classification criteria are used simultaneously to determine the species membership of an organism.

In a pragmatically oriented taxonomy, most decisions regarding species membership are made in this way. However, the problem of mixed classification lies in the fact that there is no underlying law that governs the conditions under which the use of a particular species concept is allowed versus when another concept should be employed. When is it sufficient to determine species membership only based on trait characteristics, and when is it necessary to make use of additional criteria, such as descent or gene flow relationships? There is no underlying theory that determines which species concept should be consulted in individual cases.

It is not easy to accept mixed classification, although this is the current, well-established practice. However, different species concepts are based on different biological laws. This must evoke conflicts, as documented in the monophyly-paraphyly conflict, as well as the anagenesis-cladogenesis conflict.

Evolution consists of two different processes. First, the traits of organisms change; no organism today looks as it did ten millennia ago. This process is called "anagenesis" (Figure 2.3a) (Rensch, 1947). Second, group cohesion may become lost in the course of evolution; organisms that are laterally connected to each other by gene flow will split and form separate groups. This process is called "cladogenesis" (Figure 2.3b) (Hennig, 1966).

These two processes are not directly correlated with each other. Anagenesis may occur without cladogenesis, and cladogenesis may occur without anagenesis. In the course of evolution, trait alterations can occur within a population without the population splitting up. In turn, bifurcations can take place without trait alterations (Peters, 1998). For this reason, it is difficult to unite trait alterations and cladogenetic splits with the same precision into a single taxonomic system. A decision must be made regarding which of the two processes represents speciation. It is contradictory to consider both processes as speciation, as two different evolutionary processes would then be united into one concept. It would be more consistent to avoid mixed classifications and to consider only bifurcation into separate groups as speciation and to ignore alterations of traits.

Anagenesis is a qualitative type of change: a single species exhibits alterations in its traits, whereas the number of species does not change. In contrast, cladogenesis is a numerical type of change: one species becomes two. This difference is comparable to the well-known saying of the ancient Greek naturalist Heraclitus "You cannot step into the same river twice." If you understand "the same river" to represent an identity regarding quality, then Heraclitus's saying is true. However, if you understand this to be a numerical identity, then it is always the same river because it does not have any "daughter rivers;" that is, it always remains the "stem river" over the course of time. Based on the same reasoning, a species cannot become a new species by solely by undergoing changes in its traits if it is understood as a numerical entity. A new species can only arise, if a stem species separates into two daughter species, which is a change in the number of species, not in their quality.

If a biological species is considered as an individual (not as an artificial class of trait-equivalent organisms) (Ghiselin, 1997) (Chapter 3), then anagenetic speciation has to be rejected. An individual of relationally connected organisms cannot become a new individual just because of trait alteration. Why should an individual not change its traits? It makes no sense to state that it becomes a new individual just because it has changed its traits. As an individual human being does not become a new human by changing his traits, a biological species as a unique historical product of evolution, that is, as an individual, also cannot become a new individual, that is, a new species, by changing its traits (De Sousa, 2005). The life of an individual cannot be ended by it changing its traits. The concept of the biological species as an individual forbids any anagenetic species delimitation. It is normal in the course of evolution that individuals progressively change over time and nevertheless remain one and the same thing (Ghiselin, 2002).

To avoid mixed classifications, only splits into separate groups should be termed "speciation," and alteration of traits should play no role in species delimitation. Consequently, in developing his theory, Willi Hennig did not approve of considering anagenetic change as means of taxon delimitation (Hennig, 1966). Ernst Mayr condemns this and blames Hennig for his ignorance in regard to evolution because he would not acknowledge that evolution also encompasses alteration of traits (Mayr, 1982). Mayr also recognizes speciation in anagenetic change and combines anagenesis with cladogenesis, establishing both processes as speciation. However, in this regard Mayr, cannot refer to any theoretical foundation regarding where, when and to what extent anagenesis and cladogenesis may be mixed with each other but instead refers to the intuition of the experienced taxonomist.

A very similar example of mixed classification can be seen in the monophyly-paraphyly conflict (see above). In paraphyletic systematics (as under acceptance of anagenetic speciation), two competing classification principles are combined with each other. In paraphyletic systematic, kinship relationships are mixed with classification principles according to trait equivalence. The criteria by which this combination is carried out come close to representing an arbitrary decision. Neither of the two principles, that is, grouping by kinship and classification by trait resemblance, is consistently adhered to. There is no underlying rule designating in which cases valuation of traits is allowed to take precedence over valuation of kinship relationships. There is no rule that justifies removing a particular branch from a monophylum (paraphyly), and there is likewise no scientific rule that determines in which cases it is justified to designate a species as a newly evolved species if "only" its traits have changed, without a cladistic bifurcation having occurred (anagenesis).

7.11
The Cladistic Bifurcation of a Stem Species Always Means the End of the Stem Species

The cladistic species concept is based on the strict logic of monophyly. The unavoidable consequence of this is that every cladistic bifurcation of an original stem species into two daughter species means the end of the stem species and the origin of two new species. Once one species splits into two separate groups, both groups have to be defined as new species, rather than just one of them (Figure 2.1). It would be contradictory to accept that the stem species survives while only one of the two daughter branches is designated as a new species (Hennig, 1966).

If the branch-off of a side branch was defined as a new species (*B* in Figure 2.1) while the stem species remained in existence (*A, C, E* in Figure 2.1), the bifurcation would lead to two taxa with different rankings, contradicting the principle of monophyly (Hennig, 1966). The logic of cladistics does not allow the survival of the stem species because sister taxa always must be of the same rank. From this, it follows that every bifurcation must represent the end of the stem species.

A stem species can never survive a branching event. Advocates of the view that the stem taxon continues to exist argue that the stem species has not experienced any change in its traits during the branching event from group *A* to groups *C* and *E* (Figure 2.1). Thus, only the "side branches" (*B* and *D*) would exhibit new traits and would therefore have to be considered as new species. However, making a distinction between a side branch and a main branch is an anagenetic view, and anagenesis is excluded by cladists. In cladistics, side branches as opposed to the main branch do not exist because there is no logical reason for this distinction. The logic of a side branch in contrast to a main branch inherently implies that two branches would exist with different ranks. What justification would there be for this? Only the branching is crucial, and the result of a branching event is always two new branches of equal rank, rather than only one.

Bifurcation is the disruption of gene flow. This is not associated directly with trait alteration, even though as a consequence of bifurcation, trait alterations often arise very rapidly. However, speciation is defined solely by bifurcation. The fact that trait alterations that distinguish the two daughter branches from each other and from the stem species then also occur is a secondary consequence that is not related to the definition of cladistic branching.

The awareness that speciation only means bifurcation, not necessarily the alteration of traits, is clearly counterintuitive. It is particularly difficult to accept the extreme situation: a small population at the periphery of a large geographical range becomes separated; then not only this peripheral population is considered a new species, but also the complete rest of the species. The demand that both branches of a bifurcation event must become new species, rather than just one of the two branches, requires classifying the stem species as terminated and considering its entire remainder as a newly evolved species, even if it consists of millions of individuals spread over the entire continent that did not undergo changes in any of their traits. This almost sounds paradoxical, but it is logical.

Another consequence of the concept that a bifurcation always constitutes the end of the stem species and the origin of two new species is the impossibility of determining the actual age of a species. The law of cladistics demands that the origin of a species is always based on the bifurcation of a stem species into two daughter species. There is no other mode of species formation. Consequently, the conclusion consistently arises that a species is always as old as the last bifurcation. However, as many sister branches are short lived because the organisms they represent become extinct, not all branching events are perceived by the human observer. Therefore, it is always possible that between a known split and a currently living species, another bifurcation has occurred that has not yet been detected (*B* in Figure 2.2). However, such an overlooked split would necessarily indicate that the currently living species a somewhat younger. In Figure 2.2, species *A* is younger than species *C*. As the possibility of an undiscovered lateral bifurcation can never be ruled out, the age of a species can never be determined. Thus, the question of how old a species is can never be answered. This situation is again counterintuitive, but it is logical and consistent.

7.12 The "Phylocode"

In the last ten to fifteen years, there have been efforts to abandon the hierarchical Linnaean taxonomical system and replace it with a new nomenclature, the "phylocode." The phylocode is a new taxonomic nomenclature that is intended to abandon hierarchical names (like names for genus, families, etc.) and to replace them by names for monophyletic clades.

This endeavor is linked most of all to the names Cantino, de Queiroz, Donoghue, Gauthier and Mishler, among others (http://phylonames.org/). The foundation of this approach is the realization that Linnaean taxa, as classes, can neither satisfy the claim of being real objects in nature, nor of being stable names that are valid for a longer period time. Therefore, there would be no reason at all, except for conservative clinging to entrenched thought patterns, to continue using Linnaean categories (e.g., genus, family, etc.). On the contrary, these categories hamper taxonomy due to a lack of theoretical foundations, which creates misunderstandings. Thus, the phylocode consortium aims to replace the binary Linnaean nomenclature with new phylocode names that reflect monophyletic clades.

If the phylocode does indeed replace the Linnaean system one day, it would have drastic consequences for nomenclature. For instance, the category of the genus plays a central role in the Linnaean mode of thinking and, thus, has become part of the scientific names of all organisms. The binary nomenclature of Linnaeus has survived for two and a half centuries until today, which is not difficult to understand, as it provided a significant simplification in bilateral communication compared to pre-Linnaean times. When you wanted to communicate unambiguously about certain plants and animals in pre-Linnaean times, the names were long and inconvenient. Since the time of Linnaeus, however, nobody has had to recite *Grossularia, multiplici acino: seu non spinosa hortensis rubra, seu Ribes officinarium* when he means "Redcurrant." Instead, noun designating the genus is followed by an adjective for the species, and *Ribes rubrum* makes everything clear.

However, this progress, made 250 years ago, conflicts with the laws of evolution. The misunderstanding produced by Linnaean nomenclature is the following: by the assignment of a species and genus designation, the impression is conveyed that genera of different groups of organisms are comparable to each other. After all, Linnaeus was convinced that genera existed as units in reality and so would exist even if there were no humans following the desire to sort species into convenient classes (Ereshefsky, 1999).

In contrast, the current view is that genera are nothing but human-constructed sorting units. Furthermore, genera of different animal or plant groups are not at all comparable with each other. For example, that which is referred to as a genus in more primitive animal groups represents a much higher category than a genus of birds.

Thus, if the ant genus *Formica* contains almost 300 species worldwide, but the bird genus *Aythya* (diving ducks) only includes 12 species, then via this comparison, the impression is conveyed that there is an objective difference between ants and ducks with regard to biodiversity, as if the ant genera were richer in species. However, this comparison reflects nothing more than that myrmecologists attribute species to genera more generously than do ornithologists. The genus category in different animal or plant groups does not stand on the same hierarchical level. The statement that within a particular hectare of rain forest, there are 200 genera of insects, but only 20 genera of birds thus becomes meaningless.

Yet Linnaean nomenclature is also impractical. New species are continually discovered, and existing genera therefore progressively become too large. It is impractical to work with genera that contain hundreds of species. For example, the butterfly genus *Papilio* (Swallowtails) established by Linnaeus in 1758 would contain approximately 325 species today if it had not been successively subdivided into newly established genera in post-Linnaean times (http://www.insects-online.de/frames/papilio.htm). Today, a distinction is made between nine genera of the former genus *Papilio*, such that each genus only contains 35 species on average.

Additionally, new insights regarding kinship relations appear continually, and certain species therefore must be withdrawn from one genus and categorized into another. It proves hindering that the designation of a species also contains the genus name because with every reclassification into a new genus, the name of the species has to be changed. Every name change creates communication difficulties. Among lepidopterologists (experts on butterflies), this has led, for example, to the situation that in daily practice, the currently valid genus names often are not known by field biologists and are no longer used; instead, only the species names are mentioned in everyday communications.

For these reasons, the advocates of the "phylocode" consider binary nomenclature and the Linnaean taxon hierarchy to be "out-dated" (Cantino *et al.*, 1999). The Linnaean system does not do justice to the monophyletic principle of classification because hierarchical ranks do not exist in a cladistical system. Furthermore, the Linnaean names are not especially practical, as they are too short lived.

As an alternative, the "phylocode" has been proposed with the aim of replacing Linnaean taxonomy (http://phylonames.org/; Wilson, 1999). The principle of the phylocode is a taxonomic nomenclature that names the system of living beings by consistently using monophyletic clades. The phylocode separates the entire phylogenetic tree of life into comprehensive and less comprehensive monophyla and assigns them names that no longer contain a taxonomic ranking (Cantino *et al.*, 1999). The phylocode assigns taxonomic names according to phylogenetic connections without consideration of hierarchical levels, such as genus, family, order and class. Instead, a taxon name is only assigned to bifurcating clades according to the principle of monophyly, whereby it is crucial to combine all branches that trace back to a most recent common ancestor. Although at lower levels, there are clearly clades containing only a few bifurcations, whereas at higher levels, there are superordinate clades consisting of many bifurcations, no hierarchical levels between clades are acknowledged. The increments between clades are continuous, rather than stepwise.

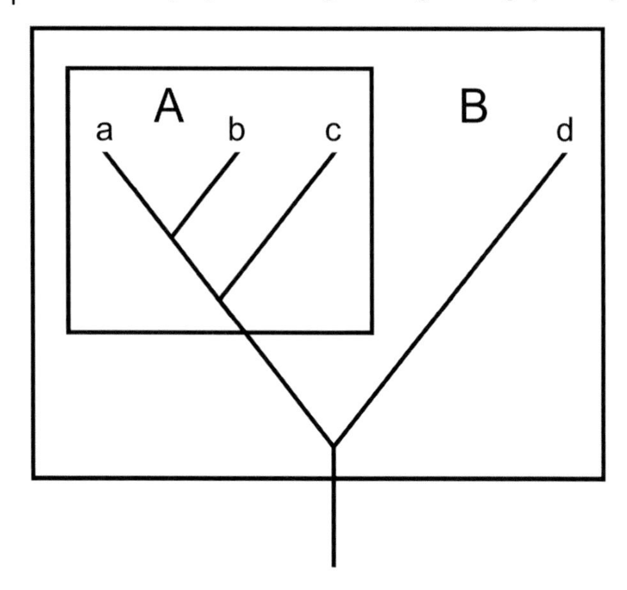

Figure 7.12 The phylocode is a taxonomic nomenclature that aims to avoid assigning names for taxonomic ranks. The phylocode taxonomy replaces the names for genus, families, and so on by names for monophyletic clades. Every phylocode taxon (e.g., *A*) is a bundle of branches (*a* through *c*) delimited by the most distant branch (*c*). Taxon *A* accordingly receives the name "Includes *c*, excludes *d*." Branch *d* then belongs to the next higher taxon *B*.

Every phylocode taxon consists of a "bundle of branches." To assign these bundles of branches names, the inclusion-exclusion principle is used. A taxon (e.g., taxon *A* in Figure 7.12) always includes one last branch bifurcating at the earliest point of time from the common ancestor moving along the time line toward the bottom of the figure. For taxon *A* in Figure 7.12, this is branch *c*. The next older branch (*d*) is then excluded from the taxon. This inclusion-exclusion principle determines a taxon's nomenclature. Taxon *A* accordingly receives the name "Includes *c*, excludes *d*." Branch *d* then belongs to the higher taxon *B*.

Thus, in the phylocode system, there are both taxa that reach far back into a phylogeny and therefore contain many bifurcations and taxa that are young and include only a few bifurcations. Hence, the sizes of the clades vary, but there are no hierarchical ranks, such as those designated as genera, families, or orders in classic systematics. A taxon always results from how far you go back in the phylogenetic tree. By going a step further back, an earlier bifurcation branch is additionally included in a taxon, and a more inclusive taxon is obtained, which contains more branches but does not assume a higher ranking.

The advantage of the phylocode system is twofold. First, the fallacy inherent in the Linnaean system that ranks actually exist and that they are comparable to each other is avoided. A researcher studying the classification of the animal and plant kingdom according to the Linnaean system must necessarily gain the impression that nature's

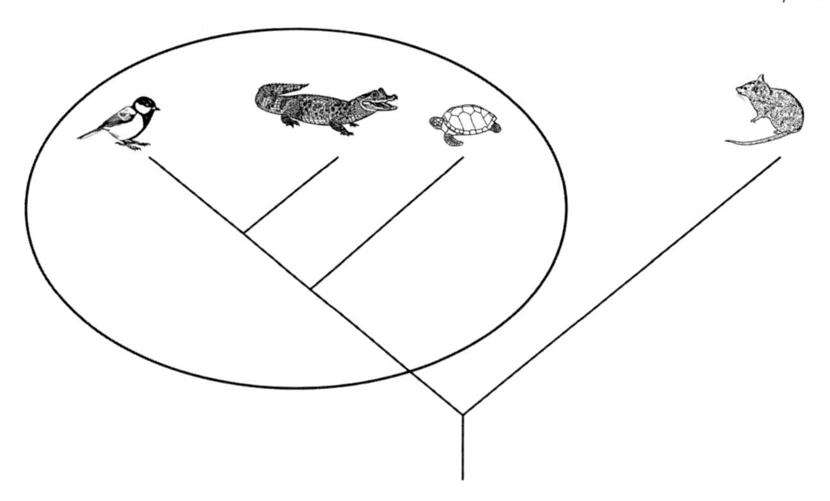

Figure 7.13 Example for the phylocode nomenclature: The name of the *Sauropsida* (reptiles including birds) would be "Includes turtles, excludes mammals."

biodiversity is divided into ranks and that, accordingly, something similar to families and orders occurs in nature. Furthermore, the impression is conveyed that these rankings are comparable to each other, whereas a mammalian family is actually not at all comparable to a nematode (roundworm) family. The second advantage of the phylocode system lies in the long-term stability of the assigned names. The names do not have to be changed constantly when new insights regarding phylogenetic relationships are obtained.

The disadvantage of the phylocode system lies in its unwieldy nomenclature. The names assigned by the phylocode system are long and cumbersome. Taxon *A* displayed in Figure 7.12 has the name "Includes c, excludes d." The group of *Sauropsida* (reptiles including birds) would have to receive the name "Includes turtles, excludes mammals" (Figure 7.13). Of course, such names would not be appropriate for everyday use. Possibly, many of the old Linnaean names will have to be retained and in a parallel nomenclature assigned to particular clades after the phylogenetic definition has been published.

The phylocode is the first system that consistently applies Darwin's theory of descent and implements it in the nomenclature. Additionally, it is the first system that attempts to overcome the incompatibility of the Linnaean system with Darwinism by completely abandoning the formation of classes.

The theoretical foundations of the phylocode system date back to the 1980s. An initial workshop meeting of the pioneers for the phylocode took place in 2002 at Yale University in New Haven, USA Here, a resolution was passed to define clade names according to the rules of the phylocode, which would then be regarded as the starting point for the specification of priorities. The first international meeting took place in 2004 at the Muséum National d'Histoire Naturelle in Paris and consisted of contributions from taxonomists from eleven countries. The "International Society for Phylogenetic Nomenclature" (ISPN) was founded at this meeting. A second meeting then followed in 2006 at the Peabody Museum of Yale University.

Up to now, the phylocode has not established itself, despite the clarity and stability of its nomenclature. The future will show whether the appeal of the phylocode can hold its ground against the hurdles of entrenched conservative thinking. Of course, the phylocode is a system of nomenclature, not an ontological solution to the species problem. The so-called "tree-thinking" on which the phylocode is based may not hide the fact that the problem of taxon delimitation is not solved by the concept of the phylocode. Taxonomy means "delimited entities." However, the phylogenetic tree is a branching continuum. There are no boundaries in the phylogenetic tree. The entities of the phylocode are "least inclusive" and "most inclusive" monophyla. The phylocode classification obeys the rule of the monophyletic formula "if *B* and *C* are combined, then *A* also belongs to this group" (Figure 7.5), but you can only combine things that are apart. Why are *A, B* and *C* actually separate branches? The phylocode does not provide any answer to this question.

8
Outlook

What is a species? The answer to this question is clear: a species is a group of individuals that are connected with each other by bonding and are separated from another group of individuals. This seems simple, but the fundamental question of taxonomy is less so: what is bonding and what is separation, and how can they be recognized and measured? Zebras and Gnus in the Tanzanian savanna are not separated. Male and female Tigers in the Indian jungle are separated for almost the entire year. Why and how are they bonded?

Cohesion of the members of a species has something to do with the (vertical) descent relationships and/or (lateral) sexual gene exchange between the organisms. There is no other consistent way to define species cohesion. What is not connected by descent or gene exchange cannot be a species because there is no natural bonding that holds individuals together.

There is an important difference between sexual cohesion and genealogical cohesion. Only sexual (lateral) gene flow can be ceased and, hence, terminate species membership. Genealogical (vertical) cohesion cannot be ceased, and therefore, it cannot not define the origin of a new species. Apart from species extinction, there is no way to define the termination of species membership via genealogical decent (vertical) cohesion because a species membership is terminated only via the interruption of the gene flow connection.

Interruption of gene flow divides a cohesively bonded group into two separate groups, resulting in speciation. Only the interruption of gene flow is speciation. In most cases, the interruption of gene flow is not complete. Most related species have occasional sexual contacts, which sometimes result in fertile offspring. Pre- and postzygotic barriers are leaky. Nevertheless, despite the existence of smooth boundaries, sexual barriers between particular groups of organisms do exist. These barriers prevent or minimize gene flow among groups, and these groups are species. Therefore, species do exist.

Finding these barriers represents a major task for the field of taxonomy. There are particular genes that are responsible for these barriers. These genes are the speciation genes; they cause pre- or postzygotic incompatibilities. Little is known about the genes that lead to mating barriers between species or to allelic incompatibilities in a species hybrid. Furthermore, little is known regarding the number of such genes and

Do Species Exist? Principles of Taxonomic Classification, First Edition. Werner Kunz.

what leads to the discrepancies in each case. This lack of knowledge is surprising because it concerns the major criteria determining why two organisms belong to different species.

Distinguishing between two objects is not the same as knowing what those objects are. It is easy to distinguish a piece of wood from a piece of iron, but distinguishing these two objects does not provide information regarding what wood is or what iron is. Wood is always a piece of a tree or a shrub and has a certain chemical composition. Iron is a metal, consisting of atoms that have a specific atomic number. Why is the question "what is" not asked more often in taxonomy? It is easy to distinguish a Tiger from a Lion, but what is a Tiger and what is a Lion? Why are they considered to be different species? This is certainly not because the Tiger has stripes, and the Lion does not have stripes. Species identification is something different from the aim of knowing why a group of organisms is a species. Two species are not two species because they are different; rather, they become different because they are two separate species. Speciation is separation, not the existence of diagnosable trait differences. Taxonomy cannot be a science that runs out into diagnosis.

Species are individuals with a temporally and spatially transient existence. Species can become extinct. Species are not classes because classes are universals that are not restricted in time or space. Classes cannot become extinct. Species cannot be "natural kinds," as there are no essential traits that would necessarily and sufficiently designate that a given individual organism has to belong to one particular species rather than to another. Changes in traits during the course of evolution (anagenesis) cannot be considered to be speciation because an individual remains the same individual for its entire life, independent of whether it experiences changes in its traits or not.

What Darwin meant with "species" in his work "The Origin of Species" were real, biological entities, not classes. A problem arises because Linnaeus also talked about "species," although he had classes in mind. Herein lies the actual root reason that no agreement has been able to be achieved regarding what a species is. Evolution does not produce biological units of organization that are well-suited to satisfying classification needs, but it was for this purpose that Linnaeus created taxa.

The biological species as an element of order and the biological species as a unit that exists in reality and plays a role in evolution are two different entities that arise from different scientific purposes. The more a species concept is used as an element of order and the more it is suited for application in practicing taxonomy, the more it is vulnerable due to lacking theoretical consistency. The reason that Linnaeus is still important in biological science is not that his view of species would be valid today but, instead, that his nomenclature retains some practical utility.

References

Abzhanov, A., Protas, M., Grant, B.R., Grant, P.R., and Tabin, C.J. (2004) Bmp4 and morphological variation of beaks in Darwin's finches. *Science*, **305**, 1462–1465.

Aguilar, A., Roemer, G., Debenham, S., Binns, M., Garcelon, D., and Wayne, R.K. (2004) High MHC diversity maintained by balancing selection in an otherwise genetically monomorphic mammal. *Proc. Natl. Acad. Sci. USA*, **101**, 3490–3494.

Ahearn, J.N. and Templeton, A. (1989) Interspecific hybrids of *Drosophila heteroneura* and *D. silvestris* I. Courtship success. *Evolution*, **43**, 347–361.

Allendorf, V.F. and Luikart, G. (2006) *Conservation and the Genetics of Populations*, Blackwell, London.

Andolfatto, P. (2001) Adaptive hitchhiking effects on genome variability. *Curr. Opin. Genet. Dev.*, **11**, 635–641.

Aquadro, C.F., Bauer DuMont, V., and Reed, F.A. (2001) Genome-wide variation in the human and fruitfly: A comparison. *Curr. Opin. Genet. Dev.*, **11**, 627–634.

Armstrong, D. (1978) *Universals and Scientific Realism*, Cambridge University Press, Cambridge.

Ast, G. (2005) The alternative genome. *Sci. Am.*, **292**, 40–47.

Atran, S. (1999) The universal primacy of generic species in folkbiological taxonomy: implications for human biological, cultural, and scientific evolution, in *Species: New Interdisciplinary Essays* (ed. R.A. Wilson), MIT Press, Cambridge, Massachusetts, pp. 231–261.

Avise, J.C., Walker, D., and Johns, G.C. (1998) Speciation durations and Pleistocene effects on vertebrate phylogeography. *Proc. Biol. Sci.*, **265**, 1707–1712.

Ax, P. (1995) *Das System der Metazoa 1: Ein Lehrbuch der Phylogenetischen Systematik*, G. Fischer, Stuttgart; Jena; New York.

Bachmann, K. (1998) Species as units of diversity: an outdated concept. *Theor. Biosci.*, **117**, 213–230.

Baldwin, M.W., Winkler, H., Organ, C.L., and Helm, B. (2010) Wing pointedness associated with migratory distance in commongarden and comparative studies of stonechats (*Saxicola torquata*). *J. Evol. Biol.*, **23**, 1050–1063.

Ballard, J.W. and Dean, M.D. (2001) The mitochondrial genome: mutation, selection and recombination. *Curr. Opin. Genet. Dev.*, **11**, 667–672.

Barraclough, T.G. and Nee, S. (2001) Phylogenetics and speciation. *Trends Ecol. Evol.*, **16**, 391–399.

Barton, N.H. (1993) Hybrid zones: Why species and subspecies. *Curr. Biol.*, **3**, 797–799.

Benzenhöfer, U. (2010) *Mengele, Hirt, Holfelder, Berner, von Verschuer, Kranz: Frankfurter Universitätsmediziner der NS-Zeit*, Klemm und Oelschläger, Münster.

Berthold, P. and Querner, U. (1981) Genetic basis of migratory behavior in European warblers. *Science*, **212**, 77–79.

Bezzel, E. and Kooiker, G. (2003) Ringeltaube: Vom scheuen Waldvogel zum Grossstädter. *Der Falke*, **50**, 5–9.

Billeter, J.C., Rideout, E.J., Dornan, A.J., and Goodwin, S.F. (2006) Control of male sexual behavior in *Drosophila* by the sex determination pathway. *Curr. Biol.*, **16**, 66–67.

Blondel, J., Dias, P.C., Ferret, P., Maistre, M., and Lambrechts, M.M. (1999) Selection-based biodiversity at a small spatial scale in a low-dispersing insular bird. *Science*, **285**, 1399–1402.

Bock, W.J. (1989) The homology concept: its philosophical foundation and practical methodology. *Zoologische Beiträge*, **32**, 327–353.

Boyd, R. (1999) Homeostasis, species, and higher taxa, in *Species: New Interdisciplinary Essays* (ed. R.A. Wilson), MIT Press, Cambridge, Massachusetts, pp. 141–185.

Bradshaw, A.D. (1972) Some of the evolutionary consequences of being a plant. *J. Evol. Biol.*, **5**, 25–47.

Breiter, M. (2008) Überlebensspezialist der Arktis - Der Eisbär: Postertier mit ungewisser Zukunft. *Biuz*, **38**, 86–93.

Brodie, E.D. (1989) Genetic correlations between morphology and antipredator behaviour in natural populations of the garter snake *Thamnophis ordinoides*. *Nature*, **342**, 542–543.

Cantino, P.D., Bryant, H.N., de Queiroz, K., Donoghue, M.J., Eriksson, T., Hillis, D.M., and Lee, M.S. (1999) Species names in phylogenetic nomenclature. *Syst. Biol.*, **48**, 790–807.

Cavalli-Sforza, F. and Cavalli-Sforza, L.L. (1994) *Verschieden und doch gleich*, Droemer Knaur, München.

Chen, X. (2002a) Continuity through revolutions. A frame-based account of conceptual change during scientific revolutions. *Philos. Sci.*, **67**, 208–223.

Chen, X. (2002b) The 'platforms' for comparing incommensurable taxonomies: a cognitive-historical analysis. *J. Gen. Phil. Sci.*, **33**, 1–22.

Christoffersen, M.L. (1995) Cladistic taxonomy, phylogenetic systematics, and evolutionary ranking (part 1). *Syst. Biol.*, **44**, 440–454.

Coyne, J.A. and Orr, H.A. (1997) "Patterns of speciation in *Drosophila*" revisited. *Evolution*, **51**, 295–303.

Coyne, J.A., Simeonidis, S., and Rooney, P. (1998) Relative paucity of genes causing inviability in hybrids between *Drosophila melanogaster* and *D. simulans*. *Genetics*, **150**, 1091–1103.

Coyne, J.A. and Orr, H.A. (1999) The evolutionary genetics of speciation, in *Evolution of Biological Diversity* (eds A.E. Magurran and R.M. May), Oxford University Press, Oxford, pp. 1–36.

Coyne, J.A. and Orr, H.A. (2004) *Speciation*, Sinauer Associates, Sunderland.

Cracraft, J. (1997) Species concepts in systematics and conservation biology - an ornithological viewpoint, in *Species: The Units of Biodiversity* (eds M.F. Claridge, H.A. Dawah, and M.R. Wilson), Chapman & Hall, London, pp. 325–339.

Crandall, K.A., Bininda-Emonds, O.R., Mace, G.M., and Wayne, R.K. (2000) Considering evolutionary processes in conservation biology. *Trends Ecol. Evol.*, **15**, 290–295.

Cronk, Q.C.B., Bateman, R.M., and Hawkins, J.A. (2002) *Developmental Genetics and Plant Evolution*, Taylor and Francis, London.

Dagan, T., rtzy-Randrup, Y., and Martin, W. (2008) Modular networks and cumulative impact of lateral transfer in prokaryote genome evolution. *Proc. Natl. Acad. Sci. USA*, **105**, 10039–10044.

Darwin, C. (1859) *On the Origin of Species by Means of Natural Selection or the Preservation of Favoured Races in the Struggle for Life*, Murray, London.

Davies, D. (2005) Atrans unnatural kinds. *Croatian J. Philos.*, **5**, 345–L 357.

Dawkins, R. (1976) *The Selfish Gene*, Oxford University Press, Oxford.

de Queiroz, K. and Donoghue, M.J. (1988) Phylogenetic sytematics and the species problem. *Cladistics*, **4**, 317–338.

de Queiroz, K. and Gauthier, J. (1990) Phylogeny as a central principle in taxonomy: Phylogenetic definitions of taxon names. *Syst. Zool.*, **39**, 307–322.

de Queiroz, K. (1998) The general lineage concept of species, species criteria, and the process of speciation: A conceptual unification and terminological recommendations, in *Endless Forms: Species and Speciation* (eds D.J. Howard and S.H. Berlocher), Oxford University Press, Oxford, England, pp. 57–75.

de Queiroz, K. (1999) The general lineage concept of species and the defining properties of the species category, in *Species: New Interdisciplinary Essays* (ed. R.A. Wilson), MIT Press, Cambridge, Massachusetts, pp. 49–89.
De Sousa, R. (2005) Biological Individuality. *Croatian J. Philos.*, **14**, 195–218.
del Hoyo, J., Elliott, A., and Sargatal, J. (1992) *Handbook of the Birds of the World -Vol. 1: Ostrich to Ducks*, Lynx Edicions, Barcelona.
del Hoyo, J., Elliott, A., and Sargatal, J. (1994) *Handbook of the Birds of the World -Vol. 2: New World Vultures to Guineafowl*, Lynx Edicions, Barcelona.
del Hoyo, J., Elliott, A., and Sargatal, J. (1996) *Handbook of the Birds of the World -Vol. 3: Hoatzin to Auks*, Lynx Edicions, Barcelona.
del Hoyo, J., Elliott, A., and Sargatal, J. (1997) *Handbook of the Birds of the World -Vol. 4: Sandgrouse to Cuckoos*, Lynx Edicions, Barcelona.
del Hoyo, J., Elliott, A., and Christie, D. (2007) *Handbook of the Birds of the World -Vol. 12: Picathartes to Tits and Chickadees*, Lynx Edicions, Barcelona.
Devitt, M. (1991) *Realism and Truth*, Princeton University Press, Princeton.
Dieckmann, U., Doebeli, M., Metz, J.A.J., and Tautz, D. (2005) *Adaptive Speciation*, Cambridge Press, Cambridge.
Dobzhansky, T. (1937) *Genetics and the Origin of Species*, Columbia University Press, New York.
Dres, M. and Mallet, J. (2002) Host races in plant-feeding insects and their importance in sympatric speciation. *Philos. Trans. R. Soc. Lond. B. Biol. Sci.*, **357**, 471–492.
Dubois, A. and Günther, R. (1982) Klepton and synklepton: two new evolutionary systematics categories in zoology. *Zool. Jb Syst.*, **109**, 290–305.
Dupre, J. (1999) On the impossibility of a monistic account of species, in *Species: New Interdisciplinary Essays* (ed. R.A. Wilson), MIT Press, Cambridge, Massachusetts, pp. 3–22.
Ebach, M.C. and Holdrege, C. (2005) More taxonomy, not DNA barcoding. *BioScience*, **55**, 822–823.
Ebert, G., Esche, T., Herrmann, R., Hofmann, A., Lussi, H.G., Nikusch, I., Speidel, W., Steiner, A., and Thiele, J. (1994) *Die Schmetterlinge Baden-Württembergs – Vol. 3: Nachtfalter I*, Eugen Ulmer, Stuttgart.
Ehrendorfer, F. (1984) Artbegriff und Artbildung in botanischer Sicht. *Z. Zool Syst. Evolut. -forsch*, **22**, 234–263.
Ehrlich, P.R. and Raven, P.H. (1969) Differentiation of populations gene flow seems to be less important in speciation than the Neo-Darwinians thought. *Science*, **165**, 1228–1232.
Enard, W. and Pääbo, S. (2004) Comparative primate genomics. *Annu. Rev. Genomics Hum. Genet.*, **5**, 351–378.
Ereshefsky, M. (1999) Species and the Linnaean hierarchy, in *Species: New Interdisciplinary Essays* (ed. R.A. Wilson) MIT Press, Cambridge, Massachusetts, pp. 285–305.
Ferguson, J.W.H. (2002) On the use of genetic divergence for identifying species. *Biol. J. Linnean. Soc.*, **75**, 509–516.
Figueroa, F., Günther, E., and Klein, J. (1988) MHC polymorphism pre-dating speciation. *Nature*, **335**, 265–267.
Fischer, E.P. (2005) *Einstein für die Westentasche*, München, Piper.
Ford, E.B. (1940) Polymorphism and taxonomy, in *The New Systematics* (ed. J. Huxley), Oxford University Press, London, pp. 493–513.
Ford, E.B. (1954) *Problems in the Evolution of Geographical Races* (eds J. Huxley, A.C. Hardy, and E.B. Ford), Evolution as a Process, London, pp. 99–108.
Frankham, R., Gilligan, D.M., Morris, D., and Briscoe, D.A. (2001) Inbreeding and extinction: effects of purging. *Conserv. Genet.*, **2**, 279–285.
Garcia-Ramos, G. and Kirkpatrick, M. (1997) Genetic models of adaptation and geneflow in peripheral populations. *Evolution*, **51**, 21–28.
Gehring, W.J. (2002) The genetic control of eye development and its implications for the evolution of the various eye-types. *Int. J. Dev. Biol.*, **46**, 65–73.
Ghiselin, M. (1974) A radical solution to the species problem. *Syst. Zool.*, **23**, 536–544.
Ghiselin, M. (1997) *Metaphysics and the Origin of Species*, State University of New York Press, New York.

Ghiselin, M. (2002) Species concepts: the basis for controversy and reconciliation. *Fish and Fisheries*, **3**, 151–160.

Glaubrecht, M. (2006) Ernst Mayr: Der "Darwin des 20. Jahrhunderts". *Der Falke*, **53**, 72–77.

Glutz von Blotzheim, U.N. (1994) *Handbuch Der Vögel Mitteleuropas – Vol. 9: Columbiformes – Piciformes*, Aula-Verlag, Wiesbaden.

Goodman, N. (1956) *The Problem of Universals*, University of Notre Dame Press, Notre Dame, Indiana.

Gornall, R.J. (2009) Practical aspects of the species concept in plants, in *Species – the Units of Biodiversity* (eds M.F. Claridge, H.A. Dawah, and M.R. Wilson), Chapman & Hall, London, pp. 171–190.

Grant, P.R. and Grant, B.R. (2002) Unpredictable evolution in a 30-year study of Darwin's finches. *Science*, **296**, 707–711.

Griffiths, P.E. (1999) Squaring the circle: natural kinds with historical essences, in *Species: New Interdisciplinary Essays* (ed. R.A. Wilson), MIT Press, Cambridge, Massachusetts, pp. 209–228.

Grobstein, C. (1979) External human fertilization. *Sci. Am.*, **240**, 57–67.

Haas, F. and Brodin, A. (2005) The Crow *Corvus corone* hybrid zone in southern Denmark and northern Germany. *Ibis*, **147**, 649–656.

Habel, J.C., Finger, A., Meyer, M., Louy, D., Zachos, F., Assmann, T., and Schmitt, T. (2009) Unprecedented long-term genetic monomorphism in an endangered relict butterfly species. *Conserv. Genet.*, **10**, 1659–1665.

Hackstein, J.H.P. (1997) Eukaryotic molecular biodiversity: systematic approaches for the assessment of symbiotic associations. *Antonie van Leeuwenhoek*, **72**, 63–76.

Haffer, J. (1982) Systematik und Taxonomie der *Larus argentatus*-Artengruppe, in *Handbuch Der Vögel Mitteleuropas* (eds U.N. Glutz von Blotzheim and K.M. Bauer), Akademische Verlagsgesellschaft, Wiesbaden, pp. 502–514.

Harvey, P.H. and Read, A.F. (1988) When incest is not best. *Nature*, **336**, 514–515.

Hebert, P.D.N., Ratnasingham, S., and deWaard, J.R. (2003) Barcoding animal life: cytochrome c oxidase subunit 1 divergences among closely related species. *Proc. R. Soc. Lond. B Biol. Sci.*, **270**, 596–599.

Hebert, P.D.N., Stoeckle, M.Y., Zemlak, T.S., and Francis, C.M. (2004) Identification of birds through DNA barcodes. *PLoS Biology*, **2**, 1657–1663.

Hebert, P.D.N., Penton, E.H., Burns, J.M., Janzen, D.H., and Hallwachs, W. (2004) Ten species in one: DNA barcoding reveals cryptic species in the neotropical skipper butterfly *Astraptes fulgerator*. *Proc. Natl. Acad. Sci. USA*, **101**, 14812–14817.

Hebert, P.D.N. and Gregory, T.R. (2005) The promise of DNA barcoding for taxonomy. *System Biol.*, **54**, 852–859.

Heine, H. (1826) *Die Harzreise*, Hoffmann und Campe, Hamburg.

Heine, H. (2006) *The Harz Journey and Selected Prosa*, Penguin Group, London.

Helm, B., Fiedler, W., and Callion, J. (2006) Movements of European Stonechats *Saxicola torquata* according to ringing recoveries. *Ardea*, **94**, 33–44.

Helm, B. (2009) Geographically distinct reproductive schedules in a changing world: Costly implications in captive Stonechats. *Integr. Comp. Biol.*, **49**, 563–579.

Hennig, W. (1966) *Phylogenetic systematics*, University of Illinois Press, Urbana.

Hey, J. (2001) The mind of the species problem (Opinion). *Trends Ecol. Evol.*, **16**, 326–329.

Hey, J. (2001) Genes, categories, and species, in *The Evolutionary and Cognitive Causes of the Species Problem*, Oxford University Press, New York.

Heywood, V.H. (1998) The species concept as a socio-cultural phenomenon - a source of the scientific dilemma. *Theor. Biosci.*, **117**, 203–212.

Hille, A. (1995) Enzymelektrophoretische Untersuchung zur genetischen Populationsstruktur und geografischen Variation im *Zygaena-transalpina*-Superspezies-Komplex (*Insecta, Lepidoptera, Zygaenidae*). *Bonner Zoologische Monographien*, **37**, 1–224.

Hirai, H. and Agatsuma, T. (1991) Triploidy in *Paragonimus westermani*. *Parasitol. Today*, **7**, 19–21.

Hollrichter, K. (2007) Species don't really mean anything in the bacterial world. *Lab Times*, **2**, 22–25.

Hull, D.L. (1968) The operational imperative: Sense and nonsense in operationism. *Syst. Zool.*, **17**, 438–457.

Hull, D.L. (1970) Contemporary systematic philosophies. *Annu. Rev. Ecol. Syst.*, **1**, 54.

Hull, D.L. (1997) The ideal species concept – and why we cant get it, in *Species – The Units of Biodiversity* (eds M.F. Claridge, H.A. Dawah, and M.R. Wilson), Chapman & Hall, London, pp. 357–380.

Hull, D.L. (1999) On the plurality of species: questioning the party line, in *Species: New Interdisciplinary Essays* (ed. R.A. Wilson), MIT Press, Cambridge, Massachusetts, pp. 23–48.

Hull, D.L. and Ruse, M. (2007) *The Cambridge Companion to the Philosophy of Biology*, Cambridge University Press.

Huxley, J. (1942) *Evolution: The Modern Synthesis*, Allen & Unwin, London.

Irwin, D.E., Bensch, S., and Price, T. (2001) Speciation in a ring. *Nature*, **409**, 333–337.

Irwin, D.E., Bensch, S., Irwin, J.H., and Price, T. (2005) Speciation by distance in a ring species. *Science*, **307**, 414–416.

Johnston, R.F. and Selander, K. (1964) House sparrows: rapid evolution of races in North America. *Science*, **144**, 548–550.

Jones, J.S., Leith, B.H., and Rawlings, P. (1977) Polymorphism in *Cepaea*: A problem with too many solutions? *Annu. Rev. Ecol. Syst.*, **8**, 109–143.

Kattmann, U. (1996) Vielfalt der Menschen aber keine Rassen. *Biuz*, **26**, 70–71.

Kermarrec, N., Roubinet, F., Apoil, P.A., and Blancher, A. (1999) Comparison of allele 0 sequences of the human and non-human primate AB0 system. *Immunogenetics*, **49**, 517–526.

Kingsley, D.M. (2009) From atoms to traits. *Sci. Am.*, **300**, 52–59.

Korol, A., Rashkovetsky, E. *et al.* (2000) Nonrandom mating in *Drosophila melanogaster* laboratory populations derived from closely adjacent ecologically contrasting slopes at "Evolution Canyon". *Proc. Natl. Acad. Sci. USA*, **97**, 12637–12642.

Kripke, S.A. (1980) *Naming and Necessity*, Basil Blackwell, Oxford.

Kunz, W. (2002) Was ist eine Art? *Biuz*, **32**, 10–19.

Lande, R. (1980) Genetic variation and phenotypic evolution during allopatric speciation. *Am. Nat.*, **116**, 463–479.

Lee, M.S.Y. and Skinner, A. (2008) Hierarchy and clade definitions in phylogenetic taxonomy. *Organ. Divers. Evol.*, **8**, 17–20.

Liebers, D., de Knijff, P., and Helbig, A.J. (2004) The herring gull complex is not a ring species. *Proc. R. Soc. Lond. B Biol. Sci.*, **271**, 893–894.

Lockwood, J.L. and Pimm, S.L. (1994) Biological diversity: Species: Would any of them be missed. *Curr. Biol.*, **4**, 455–457.

Lorenz, K. (1977) *Behind the Mirror: A Search for a Natural History of Human Knowledge*, Harcourt Brace Jovanovich, New York and London.

Magurran, A.E. (1999) Population differentiation without speciation, in *Evolution of Biological Diversity* (eds A.E. Magurran and R.M. May), Oxford University Press, Oxford, pp. 160–183.

Mahner, M. and Bunge, M. (1997) *Foundations of Biophilosophy*, Springer, Berlin.

Mallet, J. (1995) A species definition for the modern synthesis. *Trends Ecol. Evol.*, **10**, 294–299.

Mallet, J. (2004) Perspectives: Poulton, Wallace and Jordan: how discoveries in Papilio butterflies led to a new species concept 100 years ago. *Systematics and Biodiversity*, **1**, 441–452.

Mallet, J. (2006) What does *Drosophila* genetics tell us about speciation? *Trends Ecol. Evol.*, **21**, 386–393.

Mallet, J. (2010) Shift happens! Shifting balance and the evolution of diversity in warning colour and mimicry. *Ecol. Entomol.*, (Suppl 1), **35**, 90–104.

Martin, W. and Salamini, F. (2000) A meeting at the gene – Biodiversity and natural history. *EMBO Reports*, **1**, 208–210.

Mayden, R.L. (1997) A hierarchy of species concepts: the denouement in the saga of the species problem, in *Species: The Units of Biodiversity* (eds M.F. Claridge, H.A. Dawah, and M.R. Wilson), Chapman & Hall, London, pp. 381–424.

Mayr, E. (1942) *Systematics and the Origin of Species*, Columbia University Press, New York.

Mayr, E. (1963) *Animal Species and Evolution*, Harvard University Press, Cambridge, Mass.

Mayr, E. (1982) *The Growth of Biological Thought: Diversity, Evolution, and Inheritance*, Boston Mass., Havard University Press.

Mayr, E. and Ashlock, P.D. (1991) *Principles of Systematic Zoology*, McGraw Hill, New York.

Mayr, E. (1998) Two empires or three? *Proc. Natl. Acad. Sci. USA*, **95**, 9720–9723.

Mayr, E. (2000) The biological species concept, in *Species Concepts and Phylogenetic Theory: A Debate* (eds Q.D. Wheeler and R. Meier), Columbia University Press, New York, pp. 17–29.

McCarthy, E.M. (2006) *Handbook of Avian Hybrids of the World*, Oxford University Press.

McCoy, K.D. (2003) Sympatric speciation in parasites – what is sympatric? *Trends Parasitol.*, **19**, 400–404.

Menting, G. (2002) *Die kurze Geschichte des Waldes*, Mantis-Verlag, Gräfelfing.

Meyer, A. (1993) Phylogenetic relationship and evolutionary processes in cichlid fish species flocks of the East African Great Lakes. *Trends Ecol. Evol.*, **8**, 279–284.

Miller, S.E. (2007) DNA barcoding and the renaissance of taxonomy. *Proc. Natl. Acad. Sci. USA*, **104**, 4775–4776.

Mindell, D.P. and Meyer, A. (2001) Homology evolving. *Trends Ecol. Evol.*, **16**, 434–440.

Mishler, B.D. and Brandon, R.N. (1987) Individuality, pluralism and the phylogenetic species concept. *Biology and Philosophy*, **2**, 397–414.

Mishler, B.D. and Donoghue, M.J. (1994) Species concepts: A case for pluralism, in *Conceptual Issues in Evolutionary Biology* (ed. E. Sober), Mass.Inst.of Technology, Cambridge/Mass, pp. 217–232.

Mishler, B.D. (1999) Getting rid of species? in *Species: New Interdisciplinary Essays* (ed. R.A. Wilson), MIT Press, Cambridge, Massachusetts, pp. 307–315.

Moritz, C. (1994.) Defining, 'Evolutionarily Significant Units' for conservation. *Trends Ecol. Evol.*, **9**, 373–375.

Mourant, A.E., Kopec, A.C., and Domaniewska-Sobczak, K. (1976) *The Distribution of the Human Blood Groups and Other Polymorphisms*, Oxford Univ. Press, London, New York, Toronto.

Müller-Hill, B. (1999) The blood from Auschwitz and the silence of the scholars. *Hist. Phil. Life Sci.*, **21**, 331–365.

Neumann, R. (2009) How to be a species? Lab Times, 20–23.

Nichols, R. (2001) Gene trees and species trees are not the same. *Trends Ecol. Evol.*, **16**, 358–364.

Nygren, G.H., Bergström, A., and Nylin, S. (2008) Latitudinal body size clines in the butterfly Polyommatus icarus are shaped by gene-environment interactions. *J. Insect. Sci.*, **47**, 1–13.

O'Brien, S.J. and Johnson, W.E. (2007) The evolution of cats. *Sci. Am.*, **297**, 75.

Okasha, S. (2001) Why won't the group selection controversy go away? *Brit. J. Philos. Sci.*, **52**, 25–50.

Okasha, S. (2002) Darwinian metaphysics: species and the question of essentialism. *Synthese*, **131**, 191–213.

Okasha, S. (2003) Does the concept of "clade selection" make sense? *Philos. Sci.*, **70**, 739–751.

Okasha, S. (2006) *Evolution and the Levels of Selection*, Oxford University Press, Oxford.

Okasha, S. (2009) Individuals, groups, fitness and utility: Multi-level selection meets social choice theory. *Biology and Philosophy*, **24**, 561–584.

Orr, H.A. (1991) Is single-gene speciation possible? *Evolution*, **45**, 764–769.

Orr, H.A. (2001) The genetics of species differences. *Trends Ecol. Evol.*, **167**, 343–350.

Orr, H.A. (2009) Testing natural selection. *Sci. Am.*, **300**, 44–50.

Paulsen, M. and Nellen, W. (2008) Gene und Allele. *Zwei Begriffe – viele Definitionen. Biuz*, **38**, 50–55.

Paulus, H.F. and Gack, C. (1983) Untersuchungen zur Bestäubung des *Ophrys fusca* – Formenkreises in Südspanien. Probleme der Evolution bei europäischen und mediterranen Orchideen – Die Orchidee (Sonderheft), pp. 65–72.

Pääbo, S. (2001) The human genome and our view of ourselves. *Science*, **291**, 1219–1220.

Peters, D.S. (1998) On some principles of systematics. *Theor. Biosci.*, **117**, 231–236.

Phadnis, N. and Orr, H.A. (2008) A single gene causes both male sterility and segregation distortion in Drosophila hybrids. *Science*, **323**, 376–379.

Piatigorsky, J. and Kozmik, Z. (2004) Cubozoan jellyfish: an evo/devo model for eyes and other sensory systems. *Int. J. Dev. Biol.*, **48**, 719–729.

Pigliucci, M. (2002) Are ecology and evolutionary biology "soft" sciences? *Ann. Zool Fennici.*, **39**, 87–98.

Poulton, E.B. (1903) What is a species? *Proc. Entomol. Soc. Lond.*, **115**, 77–116.

Poulton, E.B. (1938) The conception of species as interbreeding communities. *Proc. Linnean Soc. Lond.*, **150**, 225–226.

Powell, J.R. (1997) *Progress and Prospects in Evolutionary Biology – The Drosophila Model*, Oxford University Press, Oxford.

Prud'homme, B., Gompel, N., Rokas, A., Kassner, V.A., Williams, T.M., Yeh, S.D., True, J.R., and Carroll, S.B. (2006) Repeated morphological evolution through cis-regulatory changes in a pleiotropic gene. *Nature*, **440**, 1050–1053.

Pulido, F., Berthold, P., Mohr, G., and Querner, U. (2001) Heritability of the timing of autumn migration in a natural bird population. *Proc. Biol. Sci.*, **268**, 953–959.

Putnam, H. (1975) *Mind, Language and Reality*, Cambridge University Press, Cambridge.

Randler, C. (2000) Wasservogelhybriden (*Anseriformes*) im westlichen Mitteleuropa: Verbreitung, Auftreten und Ursachen. *Ökologie der Vögel*, **23**, 1–105.

Randler, C. (2008) Raben- und Nebelkrähe – unterschiedliche Habitatwahl in der Hybridzone. *Der Falke*, **55**, 407.

Rensch, B. (1938) Einwirkung des Klimas bei der Ausprägung von Vogelrassen, mit besonderer Berücksichtigung der Flügelform und der Eizahl. Proc 8. Int. Ornithol. Congr. (Oxford 1934), pp. 285–311.

Rensch, B. (1947) *Neuere Probleme der Abstammungslehre*, Enke Verlag, Stuttgart.

Rensch, B. (1951) Verteilung der Tierwelt im Raum, in *Handbuch der Biologie* (ed. L. von Bertalanffy), Akad. Verlagsanstalt Athenaion, Potsdam, pp. 125–172.

Riggs, P. (1996) *Natural Kinds, Laws of Nature and Scientific Methodology*, Kluwer Academic Press, Dordrecht.

Roelke, M.E., Martenson, J.S., and O'Brien, S.J. (1993) The consequences of demographic reduction and genetic depletion in the endangered Florida panther. *Curr. Biol.*, **3**, 340–350.

Rolshausen, G., Hobson, K.A., and Schaefer, H.M. (2010) Spring arrival along a migratory divide of sympatric blackcaps (*Sylvia atricapilla*). *Oecologia*, **162**, 175–183.

Rosenberg, N.A., Pritchard, J.K., Weber, J.L., Cann, H.M., Kidd, K.K., Zhivotovsky, L.A., and Feldman, M.W. (2002) Genetic structure of human populations. *Science*, **298**, 2381–2385.

Roush, W. (1997) Biology departments restructure. *Science*, **275**, 1556–1558.

Roy, M.S., Geffen, E., Smith, D., Ostrander, E.A., and Wayne, R.K. (1994) Patterns of differentiation and hybridization in North American wolflike canids, revealed by analysis of microsatellite loci. *Mol. Biol. Evol.*, **11**, 553–570.

Saetre, G.-P., Moum, T., Bures, S., Kral, M., Adamjan, M., and Moreno, J. (1997) A sexually selected character displacement in flycatchers reinforces premating isolation. *Nature*, **387**, 589–592.

Saino, N. and Villa, S. (1992) Pair composition and reproductive success across a hybrid zone of Carrion Crows and Hooded Crows. *Auk*, **109**, 543–555.

Salvato, M. (1997) Most spectacular Batesian mimicry. Florida book of insect records, 69, 69.

Sbordoni, V., Bullini, L., Scarpelli, G., Forestiero, S., and Rampini, M. (1997) Mimicry in the burnet moth *Zygaena ephialtes*: population studies and evidence of a Batesian-Müllerian situation. *Ecol. Entomol.*, **4**, 83–93.

Schäffer, N. (2004) Schwarzkopf-Ruderente kontra Weißkopf-Ruderente: Feuer frei - im Namen des Naturschutzes? *Der Falke*, **51**, 226–231.

Scherer, S. and Hilsberg, T. (1982) Hybridisierung und Verwandtschaftsgrade innerhalb der *Anatidae*. *J. Ornithol.*, **123**, 357–380.

Schilthuizen, M. (2001) Frogs, flies, and dandelions, in *The Making of Species*, Oxford University Press, Oxford.

Schliewen, U.K., Rassmann, K., Markmann, M., Markert, J., Kocher, T., and Tautz, D. (2001) Genetic and ecological divergence of a monophyletic cichlid species pair under fully sympatric conditions in Lake Ejagham. *Cameroon. Mol. Ecol.*, **10**, 1471–1488.

Schluter, D. (2001) Ecology and the origin of species. *Trends Ecol. Evol.*, **16**, 391–399.

Schmitt, M. (2004) The species as a unit of evolution and as an element of order, in

Biosemiotik – Praktische Anwendung und Konsequenzen für die Einzelwissenschaften (ed. J. Schult), VWB – Verlag für Wissenschaft und Bildung, Berlin, pp. 79–88.

Schröer, T. and Greven, H. (1999) Verbreitung, Populationsstrukturen und Ploidiegrade von Wasserfröschen in Westfalen. *Z. Feldherpetol.*, **5**, 1–14.

Schwöppe, M., Kreuels, M., and Weber, F. (1998) Zur Frage der historisch oder ökologisch bedingten Begrenzung des Vorkommens einer waldbewohnenden, ungeflügelten Carabidenart: Translokationsexperimente unter kontrollierten Bedingungen mit *Carabus auronitens* im Münsterland. *Abh Westf Mus Naturkde Münster*, **60**, 1–77.

Seehausen, O., Teria, Y., Magalhaes, I.S., Carleton, K.L., Mrosso, H.D.J., Miyagi, R. van der Sluijs, I. *et al.* (2008.) Speciation through sensory drive in cichlid fish. *Nature*, **455**, 620–626.

Sesardic, N. (2010) Race: a social destruction of a biological concept. *Biology and Philosophy*, **25**, 143–162.

Sibley, C.G. (1997) Proteins and DNA in systematic biology. *Trends Biochem. Sci.*, **22**, 364–367.

Simpson, G.G. (1961) *Principles of Animal Taxonomy*, Columbia University Press, New York.

Smith, T.B. (1993) Disruptive selection and the genetic basis of bill size polymorphism in the African finch *Pyrenestes*. *Nature*, **363**, 618–620.

Smith, T.B., Schneider, C.J., and Holder, K. (2001) Refugial isolation versus ecological gradients. *Genetica*, **112–113**, 383–398.

Smith, T.B. (2008) Patterns of morphological and geographic variation in trophic bill morphs of the African finch *Pyrenestes*. *Biol. J. Linn. Soc. Lond.*, **41**, 381–414.

Smolin, L. (1997) *The Life of the Cosmos*, Weidenfeld & Nicolson, London.

Sneath, P.H. and Sokal, R.R. (1973) *Numerical Taxonomy: The Principles and Practice of Numerical Classification*, Freeman and Company, San Francisco.

Sober, E. (1994) Evolution, population thinking and essentialism, in *Conceptual Issues in Evolutionary Biology* (ed. E. Sober), MIT Press, Cambridge, MA, pp. 161–189.

Steinke, D. and Brede, N. (2006) DNA-Barcoding. Taxonomie des 21. *Jahrhunderts. Biuz*, **36**, 40–46.

Sterelny, K. and Griffiths, P.E. (1999) Sex and Death, in *An Introduction to Philosophy of Biology*, University of Chicago Press, Chicago.

Stiassny, M.L.J. and Meyer, A. (1999) Cichlids of the rift lakes. *Sci. Am.*, 281, 44–49.

Sutherland, W.J. (1988) The heritability of migration. *Nature*, **334**, 471.

Talbot, S.L. and Shields, G.F. (1996) Phylogeography of brown bears (*Ursus arctos*) of Alaska and paraphyly within the *Ursidae*. *Mol. Phylogenet. Evol.*, **5**, 477–494.

Tautz, D. (2009) Speciation: From Darwin to Mayr and back again. *Lab Times*, 24–27.

Tautz, J. (2008) Der Bien – ein Säugetier mit vielen Körpern. Superorganismus Bienenstaat. *Biuz*, **38**, 22–29.

Töpfer, T. (2007) Die Geschichte vom Italiensperling. *Der Falke*, **54**, 250–256.

Turelli, M., Barton, N.H., and Coyne, J.A. (2001) Theory and speciation. *Trends Ecol. Evol.*, **16**, 330–343.

van Regenmortel, M.H.V. (1997) Viral species, in *Species: The Units of Biodiversity* (eds M.F. Claridge, H.A. Dawah, and M.R. Wilson), Chapman & Hall, London, pp. 17–24.

Varga, Z.S. and Schmitt, T. (2008) Types of oreal and oreotundral disjunction in the western Palaearctic. *Biol. J. Linn. Soc. Lond.*, **93**, 415–430.

Vaurie, C. (1951) Adaptive differences between two sympatric species of nuthatches (*Sitta*). in Proceedings of the International Ornithological Congress, 19, pp. 163–166.

Venter, J.C., Adams, M.D., Myers, E.W., Li, P.W., Mural, R.J., Sutton, G.G., Smith, H.O. *et al.* (2001) The sequence of the human genome. *Science*, **291**, 1304–1351.

Verheyen, E., Salzburger, W., Snoeks, J., and Meyer, A. (2003) Origin of the superflock of cichlid fishes from lake victoria. *East Africa Sci.*, **300**, 325–329.

Vogel, G. (2001) Ecology. African elephant species splits in two. *Science*, **293**, 1414.

Wallace, A.R. (1858) On the tendency of varieties to depart indefinitely from the original type. *J. Proc. Linnean Soc.: Zoology*, **3**, 53–62.

Wayne, R.K. (1993) Molecular evolution of the dog family. *Trends Genet.*, **9**, 218–224.

Wägele, J.-W. (1995) On the information content of characters in comparative morphology and molecular sytematics. *J. Zool. Syst. Evol. Research*, **33**, 42–47.

Weitere, M., Tautz, D., Neumann, D., and Steinfartz, S. (2004) Adaptive divergence vs. environmental plasticity: tracing local genetic adaptation of metamorphosis traits in salamanders. *Mol. Ecol.*, **13**, 1665–1677.

Welch, D.M. and Meselson, M. (2000) Evidence for the evolution of bdelloid rotifers without sexual reproduction or genetic exchange. *Science*, **288**, 1211–1215.

Wickler, W. (1968) *Mimicry in Plants and Animals*, McGraw-Hill, New York.

Wilkins, J.S. (2010) What is a species? Essences and generation. *Theor. Biosci.*, **129**, 141–148.

Will, K.W., Mishler, B.D., and Wheeler, Q.D. (2005) The perils of DNA barcoding and the need for integrative taxonomy. *Syst. Biol.*, **54**, 844–851.

Wilson, R.A. (1999) *Species: New Interdisciplinary Essays*, MIT Press, Cambridge, Massachusetts.

Wirtz, P. (2000) Einseitige Hybridisierung im Tierreich. *Naturw Runds*, **53**, 172–175.

Wittgenstein, L. (1953) *Philosophical Investigations*, Blackwell, London.

Woese, C.R., Kandler, O., and Wheelis, M.L. (1990) Towards a natural system of organisms: proposal for the domains Archaea, Bacteria, and Eucarya. *Proc. Natl. Acad. Sci. USA*, **87**, 4576–4579.

Wright, S. (1935) The analysis of variance and the correlations between relatives with respect to deviations from an optimum. *J. Genetics*, **30**, 243–256.

Wu, C.I., Hollocher, H., Begun, D.J., Aquadro, C.F., Xu, Y., and Wu, M.L. (1995) Sexual isolation in *Drosophila melanogaster*: a possible case of incipient speciation. *Proc. Natl. Acad. Sci. USA*, **92**, 2519–2523.

Wu, C.I. (1996) Now blows the east wind. *Nature*, **380**, 105–107.

Wu, C.I., Johnson, N.A., and Palopoli, M.F. (1998) Haldane's rule and its legacy: why are there so many sterile males? *Trends Ecol. Evol.*, **11**, 281–284.

Scientific Terms

Alleles The multiple forms of a gene at the same genetic locus in the various organisms of a species. Each alternative form of a gene occurring at the same locus (position) on homologous chromosomes is referred to as an allele.

Allelism, multiple A series of alleles at a given gene locus. A population or species of organisms typically includes multiple alleles at each locus among various individuals (see polymorphism).

Allopatry/allopatric distribution The occurrence of two populations at separated locations. Separation prevents the organisms of the two populations from encountering each other. Allopatry is separation due to external causes (usually geographic barriers), rather than via barriers related to characteristics of the organisms themselves. Allopatrically separated populations do not need mating barriers that would prevent hybrid formation, because they do not encounter each other. Sympatric distribution is the opposite. Allopatry should not be confused with "isolation by distance," whereby cross-breeding is prevented only by distance and not by external barriers.

Anagenesis Alteration of traits of individuals of a species during the course of evolution without branching into two separate branches (Figure 2.3). Anagenesis is a qualitative transformation; the species as whole does not split into daughter species. Rather, it remains the same species, despite the change in its characters. According to some authors, however, anagenesis is considered to be a kind of speciation (see cladogenesis).

Apomorphy An alteration of a trait in the course of evolution (newly derived trait) that distinguishes the organisms of a daughter species from the organisms of the stem species (see autapomorphy).

Assortative pairing The selective choice of a mating partner based on a preference for specific traits. In most cases, the partner who makes the choice is the female (female choice). Assortative pairing is an important factor for sympatric speciation.

Do Species Exist? Principles of Taxonomic Classification, First Edition. Werner Kunz.

Autapomorphy An alteration of traits that distinguishes the organisms of two sister species from each other as well as from the organisms of the stem species after a cladistic bifurcation.

Barcoding A method of species identification, based on particular DNA sequences (so-called barcodes). Although barcoding is a method of diagnosis, it is often used misleadingly as a species concept, defining every organism as a different species that differs with respect to phylogenetic distance.

Biological species A term for a species concept that was used by Ernst Mayr as a synonym for the species as a reproductive community to make it clear that the reproductive community is not a mental construct, but an object that actually exists in nature.

Biparental reproduction The production of offspring from a zygote that arises via fusion of egg and sperm from different parental individuals (see uniparental reproduction).

Birds, migratory and sedentary Many bird species are polymorphic. Their populations consist of two different morphs, comparable to the existence of males and females within a given species. Some individuals occupy the breeding grounds only during reproduction, whereas others remain in their breeding habitats permanently. This dimorphism is genetically based.

Butterflies, uni-, bi- and multivoltine On the northern hemisphere, many butterflies include different morphs that are adapted to different geographical latitudes. In the South, bi- and multivoltine morphs dominate. Here, the butterflies produce two or more generations each year and therefore occur in quite large numbers in this region. In the North, however, butterflies of a given species hatch only once in a year. They generate only one generation and are accordingly rarer. In an overlapping region, both morphs occur side by side. This polymorphism is genetically based.

Clade Portion of the cladistic phylogenetic tree that contains at least one bifurcation (see lineage) (Figure 2.6).

Cladistics Science addressing the phylogenetic branching patterns of taxa, organisms or DNA sequences.

Cladogenesis Branching of a phylogenetic stem species into two daughter species in the course of evolution. Cladogenesis is the fragmentation of a group of cohesively connected organisms into two separate groups (Figure 2.3). Therefore, cladogenesis is speciation (see anagenesis).

Class A group of objects with equivalent traits, in contrast to a group of objects that are relationally connected which each other (Figure 2.5). A class is always a universal. It is a group of objects whose existence is not limited temporally or spatially but that can occur at any time anywhere in the world. A class is the opposite of an individual. An individual exists only once in the world. In taxonomy, the attempt to combine all organisms with similar traits into a class referred to as a species results in a group that has no real existence (see universal, individual and relational grouping).

Class, monothetic This is a class whose members all share at least one trait that must be present in all members of the class without any exceptions. If one member lacks this trait, it cannot belong to the class. The natural kind is a monothetic class (see natural kind).

Class, polythetic Class formation based on covariance of several traits. A polythetic class is characterized by a multitude (or family) of traits. A polythetic class defines unambiguous class membership, although there is no single trait that is present in each individual member of the class. Polythetic classes therefore cannot be natural kinds.

Clinal transition zone The (geographic) region where two different races of a species merge. In this zone, races produce intermediate phenotypes (see race).

Clone A group of descendants with an identical genotype produced by vegetative reproduction. This term can be applied to organisms, cells or DNA molecules.

Convergence The opposite of homology. Convergence is the evolution of similar phenotypic structures due to a common selective pressure, rather than common ancestry. Convergence is also known as "parallel evolution" (see homologous objects).

Definition A description of something that explains what something is.

Definition, operative A list of the properties of something.

Ecological niche A particular habitat occupied by a particular group of organisms. The colonization of a new niche by a group of organisms may lead to sympatric speciation.

Enhancer A DNA sequence that binds a transcription factor and thereby regulates the expression of a corresponding gene (see transcription factor).

Evolutionary systematics A mixed taxonomic classification procedure that attempts to combine two different evolutionary processes into a common, consistent grouping. Both alteration of traits (anagenesis) and splitting of a stem branch into daughter branches (cladogenesis) are considered to cause taxon formation and delimitation. This approach often creates conflicts. The best example of such a conflict is the monophyly-paraphyly conflict.

Genes Portions on the genome that occupy defined locations along the linear genome. Genes can be expressed, meaning that they encode RNA and/or proteins that determine a phenotype (see alleles).

Gene, structural A gene that encodes a protein that forms phenotypic structures.

Gene, regulatory A gene that encodes a protein that regulates the expression of other genes, for example, a transcription factor.

Gene-flow community A group of organisms that are cohesively connected via sexual gene flow. This means that the organisms or their descendants exchange genes through occasional sexual contact with each other, and their genomes recombine (Figure 2.7). Sexual lateral gene flow disturbs a dichotomous phylogenetic tree by replacing it through a network of bypasses between branches (Figure 6.1). Uniparentally reproducing organisms do not exhibit lateral gene exchange and therefore cannot form gene flow communities. The term gene-flow community must be distinguished from the term reproductive community (see reproductive community). Gene-flow community is considered to be synonymous to species.

Gene transfer (or gene flow), lateral (or horizontal) Sexual gene transfer; the transfer of genes via temporary or permanent fusion of two cells. This term includes the fusion of sperm and egg as well as conjugation in ciliates or gene transfer in bacteria or viruses.

Gene transfer (or gene flow), vertical (or genealogical) Gene transfer via reproduction; the transfer of genes from the P generation to the F generations.

Genetic drift Disappearance or increase of allelic variants without conferring any selective disadvantage or advantage. Mutation of neutral alleles is not recognized by selection because it confers no disadvantage or advantage. Nevertheless, these alleles can disappear in a population or be replaced by other alleles by chance alone.

Genotype The total set of genes of an organism (see phenotype).

Homeostatic property cluster A class of objects with traits that resemble each other. In contrast to a natural kind (see natural kind), however, the traits do not need to be present in each of the members of the class; they are not essential because they are not necessary and sufficient for class membership. It is only necessary that the traits, which lead to class membership, are sufficiently stable to be present in the members of the class with a predictable probability. The term homeostatic property class is similar to the term family resemblance.

Heterogametic and homogametic sex The heterogametic sex is the sex that possesses an X/Y, X/0 or Z/W constitution. Therefore, the heterogametic sex produces two different sorts of mature gametes during meiotic division: half are X or Z gametes, and the other half are Y, 0 or W gametes. The homogametic sex is the opposite: it produces only one type of mature gametes, X or Z gametes.

Homologous objects The descendants of replicating objects, having a recent common ancestor. Composed structures, such as complex proteins or entire organs or organisms, cannot be homologous because they are not replicating objects and do not have a common ancestor. They are assembled from components of different origins, and only some of their components can be homologous. Furthermore, the concept of homology cannot be applied to objects whose common ancestor traces far back in evolution. The term homology loses its explanatory power when applied to these objects because all life on earth ultimately has a common ancestor.

Hybridogenic speciation The origination of a new species via the hybridization of two clearly distinct parental species. Hybridogenic speciation is the origination of three species from two species. Hybridogenic speciation must be explicitly distinguished from the origination of intermediate hybrid populations in the clinal transition region of two overlapping races. Hybridogenic speciation is only possible if the hybrid individuals become immediately reproductively isolated from their parents.

Identification trait A trait that helps to identify a species, as found in field guides for plants and animals. An identification trait can only be used to identify a species that is already known. It is not possible to use an identification trait to define newly discovered organism as new species. Also DNA sequences that are used as barcodes are identification traits (see barcoding).

Individual A group of objects that exists only once in space and time, in contrast to a class, which is a universal. An individual has a beginning and an end. In contrast, a class cannot disappear as a group concept; only its members can disappear (see class).

Isolation by distance The phenomenon of the races of a species diverging distinctly when they live at some geographic distance from each other. In contrast to allopatry, however, the isolation in this case is based only on distance, not on external barriers between populations. Races that are isolated by distance may become reproductively incompatible if they are isolated sufficiently far from each other for a sufficiently long period. However, races that are isolated by distance remain connected by gene flow (Figure 6.2). Gene flow distinguishes isolation by distance from allopatry.

Lineage Segment in the phylogenetic tree between two cladistic bifurcations (see clade) (Figure 2.6).

Mixed classification The simultaneous application of more than one species concept to group organisms into a species. The logic of mixed classification is questionable. For example, if trait changes and cladistic bifurcations are both used in the determination of species membership, there is no rule by which the use of traits will have to be preferred to the use of kinship relation. The anagenesis-cladogenesis conflict and the monophyly-paraphyly conflict clearly document that the problem of mixed classification is unsolved.

Molecular clock Mutations in neutral DNA sequences accumulate roughly proportionally to the course of time. The longer the period of evolutionary divergence between two DNA sequences, the greater the number of base exchanges that will accumulate within the sequences. The number of base exchanges is used as a clock to determine the age of a branching event between two DNA sequences.

Monistic species concept The unavailing desire to find a single consistent species concept that would be applicable to all organisms. The opposite of this is the pluralistic species concept.

Monophylum A cladistic group of taxa that includes all of the descendants of a most recent common ancestor (Figure 7.5) (see paraphylum).

Morph Diagnostically distinct group of organisms within a species at the same location that interbreed without producing viable phenotypic intermediates. The best example of this is males and females (sexual dimorphism). Morphs must have different genotypes, and are thus distinguished from phenotypically different developmental stages, such as larvae in contrast to adults or seasonal di- or polymorphisms (see race).

Morpho-species A formal species concept that combines all organisms into a group that possess equal or similar morphological traits.

Natural kind A class of objects sharing equivalent traits of which at least one trait is essential, meaning that it is necessary and sufficient for class membership. If this essential trait is missing in a particular object, this object cannot member of the class, and if there is any object in the world that possesses this essential trait, this object must by definition belong to the group. Natural kinds are considered to actually exist in nature, in contrast to artificial classes, which are combinations of objects produced by human sorting efforts. The best example of natural kinds is the chemical elements, whose group affiliation is based on atomic number. In contrast, biological species are not natural kinds because there are no traits that are necessary and sufficient for species membership.

Neutral DNA sequence A DNA sequence that is not the target of selection.

Orthologous and paralogous genes Orthologs are alleles (alloforms) of a particular gene that arise during mitotic replication of the genome during mitosis. Paralogs (isoforms) of a particular gene are the result of gene duplication. All orthologous genes occupy the same chromosomal locus within the genomes of all organisms in a population, whereas paralogous genes occupy different loci in the genome. If homologous DNA sequences are aligned between different organisms, it is important to distinguish between orthologous homology and paralogous homology.

Paraphylum A cladistic group of taxa that includes only a subset of the descendants of a most recent common ancestor. Evolutionarily younger branches are excluded and classified as separate groups with the same rank as the rest of the cladistic group (Figure 7.5). The best-known example of a paraphylum is the class of reptiles, from which the subgroup of birds has been removed and designated as a separate class (Figure 7.6) (see monophylum).

Parthenogenesis Production of offspring from an unfertilized egg. Parthenogenetic reproduction is reproduction without a father. In contrast to vegetative reproduction, however, it is a type of sexual reproduction because the offspring arises from a germ cell.

Partially migratory birds A polymorphic population of birds in which the migratory and sedentary morphs of a bird species live together in an overlapping transitory region. One subset of the birds leaves the breeding habitat in fall and returns again in the spring of the following year; these are the migratory morphs. Another subset of the individuals of the same population in the same region remains in the breeding area during the winter; these are the sedentary morphs.

Phenetics A formal species concept that unites all organisms into a common species with similar traits. Phenetics is a quantitative method and is therefore independent of subjective evaluations. It is, however, an artificial system.

Phenotype Traits of an organism, ranging from direct expression products of the genes (as proteins) to complex body structures, behavioral features or developmental pathways (see genotype).

Phylocode The phylocode is an attempt to replace the outdated Linnaéan taxon names with novel names. Phylocode names are names for monophyla, including more or less branches of a cladistic tree (Figure 7.12). The phylocode aims to avoid taxonomic ranks, such as genera, families, orders or classes, because such ranks do not exist in nature, and they lead to the incorrect impression that taxa of equal ranks are comparable among different animal or plant groups.

Polygeny A single trait being controlled by many coexisting genes. A contrasting case is polypheny, in which a single gene controls several, often very different phenotypic traits.

Polymorphism This term has almost the same meaning as multiple allelism but is preferentially used for phenotypic traits, rather than for genes. Polymorphism is the entirety of the differences in the phenotypic expression of a particular gene within the individuals of a species (see multiple allelism).

Polymorphism, stable The special capacity of particular alleles to survive as multiple variants for long evolutionary period in a species without being eliminated by selection or genetic drift. Most newly originated alleles are either fixed in a population relatively rapidly or they are eliminated. In the rare case of stable polymorphism, however, the selective advantage related to the survival of the population is not based on the quality of a single particular allele, but on the existence of a multitude of alleles of a gene. This confers flexibility on a population in a changing environment because different alleles of a gene with different selective advantages are distributed among many individuals.

Population A subgroup of the organisms of a species within a delimited range of occurrence. A population is a rather artificial kind of group formation, mainly aimed at providing manageability for pragmatic purposes.

Postzygotic barrier The occurrence of hybrid incompatibility or hybrid dysgenesis in the offspring produced via species crossings. In such hybrids, the cooperation of particular genes that originate from the two different parent species, is disturbed, resulting in a reduction of fertility or vitality.

Prezygotic barrier The occurrence of morphological, physiological or ethological traits that prevent the formation of zygotes between two organisms. Prezygotic barriers prevent cross-mating between species and, hence, separate species. The origination of prezygotic barriers between two species usually requires sympatric coexistence of the two species.

Race Synonymous with subspecies. Races are groups of organisms at a geographic distance from each other that have adapted to the peculiarities of their local habitats. Different races of a species are cohesively connected to each other via gene flow in clinal transitional zones, where they produce intermediate phenotypes. Therefore, races exist only because they are isolated by distance. Races are distinguished from morphs, which coexist in the same region and interbreed in an unrestricted fashion but do not produce intermediates (see morph).

Reinforcement The origination of prezygotic mating barriers under selective pressure between two species with a syntopic occurrence if crossings between these species generate hybrids with reduced fertility or vitality. Reinforcement is often observed in species that invade the territory of other species.

Relational grouping Grouping of organisms based on criteria of relational cohesion among the organisms, for example, if the organisms are descendants of a common ancestor or members of a sexual gene-flow community. Grouping of organisms by relational cohesion is the opposite of class formation (Figure 2.5).

Reproductive community A species concept that implies that all organisms of a species actually or potentially reproduce successfully with each other. In contrast, the species concept of a gene-flow community means that all organisms that are connected stepwise via sexual gene exchange belong to a species; geographically distant organisms are not necessarily reproductively compatible with each other. Allopatrically separated organisms may belong to the same reproductive community, but never to the same gene-flow community. Organisms within a gene-flow community that are isolated by distance belong to the same species according to the concept of the gene-flow community, but may not necessarily belong to the same reproductive community (see gene-flow community).

Ring species A special form of "isolation by distance" in which geographically distant and genetically incompatible races of a species encounter each other after they have spread around an inhospitable geographic region, such as a mountain range, desert or polar region. The distribution area thus forms a ring (Figure 6.2). The peculiarity of ring species is that the distant, incompatible races of a species meet each other under natural conditions. In most cases of isolation by distance, the situation is the same, but the distant races do not encounter each other.

Self-fertilization The production of offspring from a zygote that has been fertilized by the sperm of the same parental individual that produced the egg.

Speciation, allopatric The origination of two new species under the condition that the individuals cannot encounter each other during the process of speciation; in most cases, this is because of geographic separation. The process of allopatric speciation is entirely different from the process of sympatric speciation in several respects.

Allopatric speciation proceeds without selective pressure and requires no pre- or postzygotic mating barriers, and therefore it is the result of pure coincidence

Speciation, sympatric Speciation under conditions in which individuals encounter each other regularly during the process of speciation. As sympatric speciation is the origination of two new species at the same location; this process requires prezygotic mating barriers to arise that are supported by selection.

Speciation gene A gene that is responsible for keeping two groups of organism separate from each other via preventing successful cross-breeding. Speciation genes encode the traits that are responsible for pre- or postzygotic barriers.

Species A group of organisms separated from another group of organisms via external or internal barriers that interrupt gene flow between the organisms. External barriers are usually geographic barriers (allopatry). Internal barriers are prezygotic or postzygotic incompatibilities. Phylogenetic distance alone cannot define a species because a species definition requires that the cohesive connection between the members of two species is broken.

Species, cryptic Two or more different species with (almost) identical traits.

Species, polytypic A species consisting of different morphs.

Species definition The ontological definition of a species. The criteria determining how two species can be distinguished from each other are not sufficient to act as a species definition.

Species identification The procedure for distinguishing two species from each other on the basis of particular identification traits. Species identification is a diagnostic procedure and is not the same as species definition.

Stem species A species in the cladistic tree prior to its division into two daughter species.

Subspecies Synonymous with race.

Sympatry/sympatric distribution The occurrence of two populations that encounter each other regularly at the same location. Sympatric distribution is only possible if the individuals of the two populations possess mating barriers that prevent hybrid formation (see pre- and postzygotic barriers). Allopatric distribution is the opposite.

Taxon A group of organisms that is useful for taxonomic classification.

Taxonomy The science of combining organisms into groups.

Transcription factor A protein that regulates the time and intensity of the expression of a corresponding gene by binding to an enhancer (see enhancer).

Uniparental reproduction The production of offspring from only one parent. This can occur via self-fertilization, parthenogenesis or vegetative reproduction. There is no sexual contact between males and females under this scenario.

Universal Universals are general qualities that a group of individuals or particulars share, for example, a certain color or a number. The problem of universals is about their status; as to whether universals exist independently of people's minds. For example, class formation in taxonomy arises from attempts to account for the phenomenon of similarity among particular organisms (see class). Tigers in East Asia are similar, namely in having several attributes in common. The issue, however, is how to account for this fact. The problem of universals is about whether universals as such exist. Many philosophers agree that some universals exist (natural classes or natural kinds), while other universals are nothing else than names made by humans (artificial classes). Both types of universals, however, are not spatio-temporally restricted, and as such they are the opposite of an individual (see class and individual). Galaxies, for example, are a universal because they exist repeatedly in the world. In contrast, the Milky Way is an individual, because it exists only once.

Vegetative reproduction The production of offspring from somatic cells. Vegetative reproduction is reproduction without germ cells and without sexual contact (see clone).

Index

Do Species Exist? Principles of Taxonomic Classification, First Edition. Werner Kunz.